SOME READERS' REVIEWS

After months of pain and misery and bucket loads of painkillers this book lifted me out of the hole. Within days of performing Sarah's exercises I was back to work and enjoying a comfortable night's sleep. Three weeks later and it is hard to imagine I have been diagnosed with a prolapsed disc. My recovery is nearly complete. If it had not been for this book I would eventually have had to gamble with surgery. Even my GP is surprised by the turnaround. Many thanks to Sarah Key.
February 2007

I suffer from back pain. This book has been of great help to me. The early chapters describe the back and explain in very understandable terms what's going on. Trying to pin down exactly what condition you suffer from can be a bit tricky, and I would suggest you take the book with you to your physio. However, I have found this is not crucial. Buy the book, use it in conjunction with professional help, but get busy with the exercises and you'll believe in your back again.
January 2007

I have been suffering with chronic back pain now for 6 years and have done the rounds of the medical profession to try to come to some sort of diagnosis so that I can get to work on rehab. Unfortunately all of what I have been told and all that has been physically done to me simply have not worked. That was until I discovered Sarah Key's book. It totally opened my eyes to the reality of what exactly was and IS going on with my spine. That mental picture is of great benefit alone, just to be able at last to understand the spinal structure and what exactly CAN go wrong and how it CAN be corrected. This book and Sarah's methods are by far the best I have come across in all my time of back pain. GO BUY IT!!!!
January 2006

Having had a 'bad back' on and off for years, a recent very bad patch made me think it was time to find out what might be causing it, and if there was anything I could do about it. I found this book very informative. I have been doing the exercises described for a few weeks now and feel a substantial (and hopefully permanent) improvement. I like to understand what exercises are supposed to achieve and how—this book has explained both aspects.
September 2005

I couldn't put the book down! Loads of information, that is useful and helpful for my back pain. Chapters are detailed and easy to read, not over the top on medical terms. I recommend this book.
August 2005

I bought this book for my mum. She has suffered for 20 years with back pain and takes anti-inflammatory drugs daily. My mum is 58 and I just hate to watch her suffer more and more every year and I am frustrated that I can't help her. Getting her this book is just another book in a long line of self-help and exercise books that I have bought for her over the years. But this book really seems to have made some difference! She has told me again and again how great this book is because she now understands for the first time what is wrong with her back. She has also regained hope that her back can actually be treated! I am so pleased to hear her say that because I know she had almost given up. So thank you so much.
July 2005

I have just purchased Sarah Key's *Back Sufferers' Bible* and I have to say it is absolutely brilliant. I was in absolute agony with my bulging disc and was given one exercise to do from a physiotherapist that did not work. I have now done the two exercises that Sarah suggested for my problem and I am now so much more comfortable and able to manage my back pain. I cannot recommend this book enough. It was hard to believe that the two easy exercises I did would work but they did. Thank god for Sarah Keys.
August 2003

I just wanted to say thank you. I have suffered from minor back problems for years. Last year it got a lot worse and I spent many painful days getting different opinions from doctors and trying to understand my problems. When I finally 'discovered' this book it was like a breath of fresh air. I was recommended for surgery due to sciatica and loss of feeling in my left leg, but managed to avoid it. Now I am steadily getting better via self-help.
July 2003

Well written and detailed. The commonest causes of back pain are explained. Excellent exercises are well described. Useful for professionals and self-help alike.
August 2002

SARAH KEY'S

BACK SUFFERERS' BIBLE

Vermilion
LONDON

25 27 29 30 28 26

This edition published in 2007 by Vermilion,
an imprint of Ebury Publishing
A Random House Group company

This edition first published in Australia by Allen & Unwin in 2007

The Random House Group Limited Reg. No. 954009

Addresses for companies within the Random House Group can be found at
www.randomhouse.co.uk

A CIP catalogue record for this book is available from the British Library

Penguin Random House is committed to a sustainable future for
our business, our readers and our planet. This book is made from
Forest Stewardship Council® certified paper.

Printed and bound in Great Britain by Clays Ltd, Elcograf S.p.A.

ISBN 9780091814946

Copies are available at special rates for bulk orders. Contact the sales development
team on 020 7840 8487 for more information.

To buy books by your favourite authors and register for offers, visit
www.randomhouse.co.uk

Foreword
By HRH The Prince of Wales

ST. JAMES'S PALACE

Anyone writing about anything complicated knows how hard it is to hit the right note; to convey the right amount of information without lapsing into technical jargon so you leave your reading public behind and, at the same time, not dumbing it all down.

In this new book Sarah Key proposes, in readable language, a new framework for the way healthy spines break down through different stages and become painful. In a classic case of not seeing the wood for the trees, she believes that most spinal problems start off as a simple stiff link in the spine but it is so low-tech it escapes the attention of conventional medical 'wisdom'. Despite backpain being universally commonplace, she says it is often ignored until something more serious develops (like arthritis of the spine or a slipped disc) which is more easily recognised and diagnosed.

I find Sarah Key's book makes sense in what to many is a pretty incomprehensible subject. It is such a help to have it explained so one can actually understand what is going on. Visualising what is happening inside the back makes it much more logical and easy to see why Sarah Key's exercises really do work. After all, I should know. As one of her guinea pigs over the years I can vouch for their effectiveness, if not claim some credit for honing the final product.

To Anne Kern, my editor, whose enthusiasm has been my well-spring.

Contents

Foreword   v

Introduction   **1**
The way it might be   1
The way it is . . .   2
The way it goes wrong   5
The way ahead . . .   10

1 How a normal spine works   **13**
What is a spine?   13
    The lumbar vertebrae   15
    The spinal ligaments   19
    The intervertebral discs   23
    The nutrition of the discs   28
    The vertebral movements   30
    The facet joints   33
    The bending human spine   36
    The spinal nerves   40
    The muscles which work the spine   41

2 The stiff spinal segment   **50**
What is a stiff spinal segment?   50
Causes of a stiff spinal segment   52
    Unremitting spinal compression reduces disc metabolism   52
    Gravity squeezes fluid from the discs   53

Abdominal (tummy) weakness allows the spine to 'sink'   55
Sustained postures accelerate fluid loss and poor varieties of
movement prevent fluid replacement   56
Chronic protective muscle spasm compresses the problem disc   62
Abnormal postures increase neurocentral compression and
reduce metabolic activity of the discs   64
Injury can rupture the cartilage endplate between vertebra
and disc   69
The disc breaks down   71
The way this back behaves   74
   The sub-clinical phase   74
   The acute phase   74
   Acute palpation   76
   What causes the acute pain?   76
   The sub-acute phase   77
   Sub-acute palpation   78
   The chronic phase   78
   Chronic palpation   78
   What causes the chronic pain?   79
What you can do about it   80
   Aims of self-treatment for segmental stiffness   80
   A typical self-treatment for acute segmental stiffness   81
   A typical self-treatment for sub-acute segmental stiffness   82
   A typical self-treatment for chronic segmental stiffness   83

3   Facet joint arthropathy   85
What is facet joint arthropathy?   85
   Diagnosis by manual palpation   89
Causes of facet arthropathy   91
   Disc stiffening allows the facet capsules to tighten   91
   Disc narrowing causes the facet joint surfaces to override   92
   A sway back causes the lower facets to jam   94
   Weak tummy muscles can jam the facets   97
   A shorter leg invokes a greater restraint role of the facets   98
   Golf clinic   101
The way this back behaves   101
   The acute phase   101

Manual diagnosis of an acutely inflamed facet   102

What causes the acute pain?   103

Interrupting the pain cycle   104

The chronic phase   105

Lessening the pain   105

What causes the chronic pain?   107

What you can do about it   109

The aims of self-treatment for facet joint arthropathy   109

A typical self-treatment for acute facet joint arthropathy   110

A typical self-treatment for sub-acute facet joint arthropathy   111

A typical self-treatment for chronic facet joint arthropathy   112

4   **The acute locked back**   **114**

What is an acute locked back?   114

Causes of an acute locked back   117

A natural 'window of weakness' early in a bend   117

Segmental stiffness predisposes to facet locking   119

Muscle weakness contributes to facet locking   120

The way this back behaves   122

The acute phase   122

What causes the acute pain?   123

The sub-acute phase   124

The chronic phase   125

What you can do about it   126

The aims of self-treatment for an acute locked back   126

A typical self-treatment for acute locked back   127

A typical self-treatment for sub-acute locked back   128

A typical self-treatment for chronic locked back   129

5   **The prolapsed 'slipped' intervertebral disc**   **132**

What is a prolapsed disc?   132

Diagnostic techniques   137

Disc surgery   139

Causes of a prolapsed disc   141

Pre-existing breakdown alters the properties of the nucleus and weakens the disc wall   142

*Bending and lifting stress breaks down the back wall
of the disc* 143
Intensifying the breakdown 144
The way this back behaves 146
The acute phase 146
What causes the acute pain? 148
The chronic phase 150
What causes the chronic pain? 153
What you can do about it 154
The aims of self-treatment of a prolapsed disc 154
A typical self-treatment for acute prolapsed disc 155
A typical self-treatment for sub-acute prolapsed disc 156
A typical self-treatment for chronic prolapsed disc 157

6 The unstable spinal segment 158
What is segmental instability? 158
Diagnosis 162
Spinal surgery 164
Causes of segmental instability 165
Primary breakdown of the disc 165
Primary breakdown of the facet joints 168
Incompetence of the 'bony catch' mechanism of the facet joints 169
Weakness and poor coordination of the trunk muscles 171
Some speculation 172
The way this back behaves 173
The acute phase 173
What causes the acute pain? 174
The sub-acute phase 174
The chronic phase 176
What causes the chronic pain? 178
What you can do about it 178
The aims of self-treatment for segmental instability 178
A typical self-treatment for acute instability 179
A typical self-treatment for sub-acute instability 180
A typical self-treatment for chronic instability 181

7   Treating your own back   **183**

Helping yourself   183
   Preliminary thoughts   184
The procedures   186
   Bed rest   186
   Medication   188
   Painkillers and NSAIDs   189
   Muscle relaxants   190
Exercises for treating a bad back   192
   Rocking the knees to your chest   192
   Rolling along the spine   195
   Legs passing   198
   Reverse curl ups   199
   The BackBlock   202
   Segmental pelvic bridging   208
   The Ma Roller   210
   Squatting   212
   Toe touches in the standing position   214
   Diagonal toe touches   216
   Floor twists   218
   The Cobra   219
   The Sphinx   222
   Spinal intrinsics strengthening   223

Reference reading   **227**

Introduction

THE WAY IT MIGHT BE . . .

There is something sublimely beautiful in the way the human body moves which is often played out in sport. Something about the hum-drum splendour of each 'grand action', like a workaday form of ballet: the golfer leaning sideways into his swing and then winding up to the finish, his trunk twirled and elbows held high. Or the swimmer carving a furrow through the water, his body rolling languidly behind as his legs make their slow-motion, thin scissors kick.

I wonder too, whether some sporting actions affect us because they strike a deeper chord, echoing some long past function in our prim-itive memory; actions innate but mostly forgotten. The spread-eagled star shape of the javelin thrower, at the point of letting it fly. Or the horse and its rider, both flattened out at the gallop or, better still, the curling bursting thrust of a spine unravelling backwards as the rower pulls his scull across the water.

To me, rowing is the most beautiful sporting action of all. And smitten as I am, I wonder if the instinct has been in my blood all along; some tenuous calling like a wispy, elusive thread through history, linking me to an earlier image of sunlight flashing on a flank of oars. Perhaps this fleeting resonance with the days of Phoenician galleys and Viking longboats has caused me—in my fifth decade—to embark on my own odyssey by learning to row.

In that semi-transcendent state of synchronised exertion, as my white-gloved hands describe neat semicircles in to my ribs and I hear the muted clonk of the riggers as all eight blades turn and feather as one, I feel I am in touch with something deeper than the mere thrill of bubbles bursting against the hull and the sun peeping over the headland, its first rays dancing in shards on the dimpled water.

With sweat trickling down my brow, in awe I watch the back of my crew-mate in front of me—because you never see a spine working better than that. As she curls forward, hands stretched over the gunwales, legs loaded up at the catch, she is about to roll back in one beautifully timed flowing action where the power of her legs and trunk straightening, as her arms finish the stroke, adds up to a sum of energy greater than its parts.

For someone who has spent the whole of her professional life watching spines that can barely bend forward to pick up the tooth-paste, I sense another confluence on why it is that things come to pass.

THE WAY IT IS . . .

Back pain is on the increase. There is probably not a soul on Earth who has not been troubled at some stage by it, or not known someone else who has. Because of our lifestyle, back pain is more widespread now than ever before. Universal automation has caused us to go soft and our spines struggle to cope with long periods of indolence punctuated by jarring over-exertion.

I suspect the origins of back pain are simpler than we ever dreamt possible: a benign 'linkage' problem caused by the stiffening of a disc–vertebra union at the front of the spine. As the disc dries and gets harder and the vertebra on top loses mobility, the segment becomes sluggish, like a stiff link in a bicycle chain. This is then the 'derivative spinal condition' and the starting point from which other more serious problems can flow. The ensuing cascade of breakdown is the central theme of this book.

All of us are walking around with spines riddled with stiff links, never knowing they are there. They simply lie there, like sleepers, and rarely come to light. But in some cases—particularly in the lower spine which carries more weight—the link can become so stiff it becomes painful. And this, I believe, is the chief cause of common or garden backache.

People who use their hands to alleviate backache—such as physiotherapists, chiropractors, osteopaths and to some extent masseurs—can feel a painfully jammed segment, like a plug of cement in a rubber hose. Probing about with the thumbs or heel of the hand we can feel if a vertebra does not yield or is out of alignment; if it doesn't go, it doesn't go.

Conventional orthopaedics has never recognised 'segmental stiffness' as a subliminal spinal disorder—far less felt about for it with the hands—which makes it abundantly clear why our two strands of medicine to this day remain so divergent. The nearest we get is the diagnosis of 'internal disc disruption' although this takes no account of the many lesser forms of the condition which fail to warrant the highly invasive 'positive discography' diagnostic testing to confirm.

Manual practitioners who interface with the back-suffering public in droves are constantly unearthing stiff and painful vertebrae in spines. These are what we work on all day, despite their evading positive identification with any number of imaging techniques. Finding stiff links in the spine and prising them free is central to our way of working. Yet this entity, this thing, that occupies us so, has neither name nor place in the medical lexicon. What we work on all day is not described in the swathes of medical literature. Too low-tech to feature on the radar; too simple for words.

I believe the medical profession has focused 'too deep and too narrow' on backs, looking for cold hard evidence of anomalous spinal structures—enlarged or broken, say—instead of searching and listening in a more subtle way for function faults which may be reversible. Looking only for the whizz-bang stuff can shift the emphasis so far awry that evidence unable to be found presents another quandary: the sufferer who is disbelieved or, worse still, dismissed as malingering or making a fuss.

In the realm of back surgery the emphasis has been on removing bits and pieces of disc, boring holes to give nerves more room, or joining one vertebra to another, all in themselves quite drastic measures.

Modern techniques of inserting artificial discs are equally invasive (and unproven over the longer term) though the newer advances of autologous disc cell grafting (inserting cells from your own healthy discs into the sicker one) is more hopeful because it incorporates the regenerative potential of ageing discs.

In the world of backs, the interest is still focused firmly on surgical intervention while ignoring the proposition that discs can repair, regenerate or heal. In fact, in some circles this notion is dismissed out of hand. My own working hypothesis drawn from three decades of professional life is that—given the chance—discs *do* repair and that simple techniques and procedures to encourage this are the central thread of spinal therapeutics.

It is, however, a process and does not happen overnight. Discs have a slow metabolic rate, so they regenerate slowly (just as they degenerate slowly) but they are not lifeless structures. You will read later how breakdown is primarily caused by nutritional interference and sustained activities subjecting discs to sustained loading which reduces their metabolic rate even more.

This should be good news to sufferers because it shines a light, no matter how flickeringly feeble, into a corner where patients can take themselves and quietly get to work. Jammed links in the spine often need preparatory loosening by a therapist to make the stiff segment freer to suck and blow with the rest of the spine, but most of the slow and often painful process can only be executed by you.

It is significant that segmental stiffness is impossible to pick up without using hands. It never shows up with conventional imaging of X-rays and CT scans, any more than a photograph shows up a stiff hinge in a door. However, recent magnetic resonance imaging (MRI) developments have been very promising and provide the first glimmer of evidence to corroborate what people like me can feel with our hands. With the quality improving all the time, MRIs can now calculate reduced water content (from which it takes its signal) with this 'dryness' equating to the stiffness we therapists can feel.

The fact remains: back pain is pandemic. It permeates all nationalities, all social groups and all professions, and in modern times is second only to the common cold as a cause of time off work. There are literally millions upon millions of sufferers in the world today no closer to having their questions answered or their problem solved. A relatively recent survey carried out in Britain discovered 'a profound and widespread dissatisfaction with what is at present available to help people who suffer from back pain' (Department of Health and Social Security, 1996). Likewise, in the United States 85 per cent of the people who visit their medical practitioner 'leave with no nuts and bolts reason for their pain' (*Scientific American*, August 1998). At the beginning of the new millennium it seems we are little further ahead. In a manner of speaking, we are still chasing our tails.

Over the years various syndromes have taken the blame for back pain. The most enduring has been the so-called 'slipped' disc but others have taken their share: arthritis, joint sprain, muscle spasm, muscle tear, pinched nerve, blood clot, twisted sacrum. The list is endless and invariably it is none of the above.

Against all odds, I find myself roused to hypothesise a new model for the breakdown of spines, simply because the present one is so lacking. To date, little has been ventured in mapping a sequence of decay from commonplace simple backaches, to the complex incurable ones. The conventional view has always seen spinal problems popping up at random—with no relationship to a pre-existing more benign disorder, and no part to play in creating a more serious one down the track. Getting a handle on the 'treatment' has been just as ramshackle and, consequently, a riotous variety of back therapies abound—from drastic surgery at one end to realigning the pull of muscles or off-loading suppressed birth traumas at the other.

In parts of this book I have made intuitive leaps of faith in explaining how the spine works, and how things deteriorate when a simple fault goes unchecked. You just have to bear with me; I am trying to establish links between known spinal mechanics and the broken-down wrecks I see before me every day.

THE WAY IT GOES WRONG . . .

I propose that a simple back pain develops when an intervertebral disc (the fibrous pillow between the vertebrae) loses water and stiffens. This can be caused by several factors, not least small-scale injury—

either across or through the length of the spine—and long-term compression. Then one of two things can happen for the problem to get worse: you can develop more serious trouble from the front compartment of the spine as the disc breaks down further. Or you can develop trouble from the back compartment as strain translates across the spine to the facet joints. Worse still, you can get pain from both compartments at the same time. Finally, the wholesale destruction of both compartments can cause the vertebrae to slip around, in what is called segmental instability.

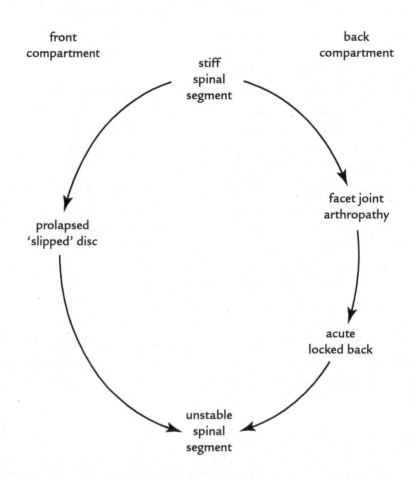

Diagram 1 A spinal segment can break down via the front compartment as the disc stiffens, or through the facets of the back compartment. As the two structures weaken the segment may become 'unstable'.

The sequential disorders in the route of breakdown are as follows:

Stage 1: A stiff spinal segment

An intervertebral disc ceases to be a buoyant pillow and becomes like a hard non-compliant washer between two vertebrae. The vertebra on top loses mobility and the segment becomes like a stiff link in the spinal chain. Because the disc lacks movement it cannot suck and blow to feed itself, so it shrinks all the more and further retires from activity. Eventually the flattening can be picked up on X-ray, but well before this point the condition can be painful—in what I believe is the most common spinal disorder.

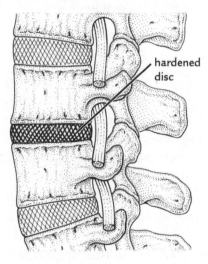

hardened disc

Figure 1 A stiff spinal segment is like a stiff link in a bicycle chain.

Stage 2: Facet joint arthropathy

As a disc at the front of the spine drops in height it causes overriding of the bone-to-bone junctions (called the facet joints) at the back of the spine. The upper vertebra settles down on the one below, causing bony rub between parts of the spine which should only have fleeting contact. Early on, this simply inflames the soft tissues around the facet joints but eventually it causes arthritic change as it erodes the cartilage covering the bone. Facet joint trouble also is a relatively common form of low-back pain.

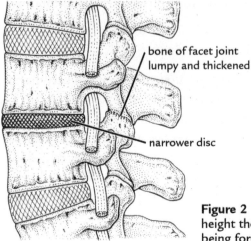

bone of facet joint
lumpy and thickened

narrower disc

Figure 2 As the interceding disc loses height the facets at the back suffer by being forced to bear more load.

Stage 3: The acute locked back

This is a fluke incident when you are caught off-guard by a movement with pain like an electric cleaver going through your back. The body locks itself rigid and moving in any direction is excruciating. Although there never appears to be any warning, the problem usually has its origins in lack of hydrostatic pressure caused by incipient disc break-down and corresponding weakness of the local deep spinal muscles.

All spines, even healthy ones, must brace themselves as they pass through a vulnerable part of range at the beginning of a bending move-ment. If a disc between two vertebrae has flattened through the degenerative process, it may be unable to generate sufficient springing-apart tension to keep its segments stable as the spine goes over. The top vertebra can imperceptibly slip askew or mis-joint at one of the facets, and the muscles go into instantaneous protective spasm to stop the slippage going further.

Stage 4: The prolapsed 'slipped' disc

As a disc progressively degenerates, the ball of fluid at the centre (the nucleus) dries out and load is transferred to the disc walls. In some cases the wall perishes at the points of greatest duress—usually one of the back corners. As the nucleus dries it also loses cohesion and with

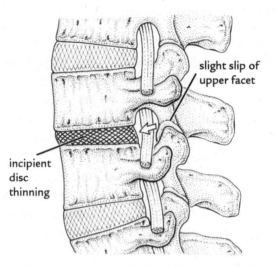

Figure 3 A lower intradiscal (hydrostatic) pressure can fail to 'prime' the disc, so that the vertebra above can slip minutely askew at a facet joint as the spine goes to bend.

excessive twisting and lifting activity it can extrude through fissures in the wall where it is the most broken up. The pressure of the runaway nucleus against the sensitive outer layers of the disc wall can cause severe back pain. Sciatica (leg pain) can also be caused by the displaced nucleus lodging on a nearby spinal nerve root.

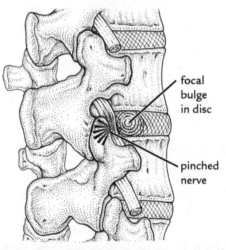

Figure 4 Disc prolapse is caused by a degenerating nucleus squeezing through a fissure in the disc wall, to be retained by the outer tensile layers of the wall.

Stage 5: The unstable spinal segment

With progressive loss of internal pressure, the disc cannot spring-load its vertebra when the spine bends. With each movement, it goes to shear forward at the problem link, tugging at its own walls as it goes. As it perishes, more strain is taken by the other main structure holding the segment together, the capsular ligaments of the facet joints at the same level. Eventually these ligaments stretch too, leaving the vertebra to wobble about in the column.

In the event of severe arthritic change of the facets, instability can spread from the other side of the segment first. Eventually the disc suffers because stretched facet capsules allow too much movement of the segment. Frank instability of a segment is not a common cause of back pain.

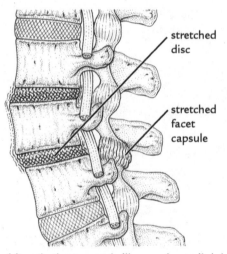

stretched disc

stretched facet capsule

Figure 5 An unstable spinal segment is like one loose link in the spinal chain.

THE WAY AHEAD . . .

The good news is that the right therapy almost anywhere along the route of spinal breakdown can stop it in its tracks and turn it around. Often curing a problem employs in reverse the same principles of destruction which brought it about in the first place. The linear progression through increasingly complex syndromes is just as straightforward in reverse.

The other good news is that you can do most of the rehabilitation yourself. Phase by less-painful phase you can steer yourself back out of the maze where you have been stumbling about for so long. At first you will barely believe it, or think you must be imagining things. Then in countless small ways you will feel your load lightening and your movements quickening. As you pass the hall table, you will pick up the newspaper on the run instead of planning every move. At last you will savour the sweet thrill of hope.

True, the passage of recovery may be far from smooth, and this is spelled out very clearly in the self-help sections of the book. The chapters on the various disorders contain suggested treatment regimens for the different phases, although you will feel free to vary your own regimens slightly once you become au fait with your problem. You will find you undergo occasional (and inevitable) hiccups, even under the control of a therapist, which should make you less anxious if this happens to you.

Knowing how the spine works is critical, and for this reason there is quite a lot of detail in the book about spinal function. Seeing how it goes wrong is equally important, and by describing the symptoms of each disorder (as much as I specifically can) you are helped to know what your problem is. I also describe what I feel when I delve around in a back, and though this is not strictly relevant from your point of view, it throws more light on the wider picture. Remember, everything is much easier when you understand. Understanding is half the cure.

This book is designed to rescue sufferers from a wilderness of conflicting opinions and advice. However, there is no substitute for early hands-on treatment from a qualified therapist, preceded by careful diagnostic screening from a medical practitioner to exclude anything sinister. Your therapist will isolate the problem level, after which he or she will manually mobilise it to get the ball rolling. Then you will follow on with the part of treatment that only you can do.

Knowing when to advance and when to backpedal on treatment is an ill-defined area which is discussed in detail. Even therapists find this difficult to gauge, and except for a specific knack using the hands, it is probably the one area which differentiates levels of skill. On the other hand, you on the 'inside' have instincts when treating yourself, which are powerful allies. Keeping calm and in touch with your gut feelings, without being excessively introspective, will always steer you through a rough patch.

Ultimately, self-treatment puts you in charge by making you responsible for your own back. It wrests you free of the endless rounds of appointments—going from one practitioner to the next—and takes you out of the passive role of being a patient. It keeps you improving without being dependent, which means you are no longer the victim. Better still, it makes you the architect of your own destiny. It means . . . it's up to you!

1

How a normal spine works

This chapter describes the nuts and bolts of how the spine works. Parts of it are quite technical, particularly the mechanics of bending and the function of the various muscles, but I'm afraid there is no way around this. It is simply not possible to understand how things go wrong without first seeing how they should work. More to the point, the information lets you know what you are doing when it comes to fixing your back.

WHAT IS A SPINE?

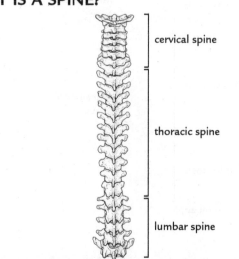

Figure 1.1 The human spine is a slender segmented column made up of 24 segments sitting atop the narrow sacral base.

The human spine is an upright bendy column. It consists of 24 separate segments called vertebrae which sit atop each other in a vertical stack. There are seven in the neck (cervical), twelve in the middle back (thoracic) and five in the low back (lumbar). The base of the spine sits on the sacrum, which is a solid triangular block of bone at the back of the pelvis. The sacrum tilts down at the front to an angle of approximately 50 degrees below the horizontal, making a concavity in the low back as the spine arches to compensate.

The spine rises out of the pelvis in three gentle curves like a cobra from a basket. Its 'S' shape helps hold it upright, and by arching back and forth over a central line of gravity it balances the top-heavy torso over its narrow base. With perfect spinal alignment (posture) a straight line can be drawn through the ear, the tip of the shoulder, the spine at waist level, the front of the knee joint and the back of the ankle.

The hollow in the low back is called a lumbar lordosis. This is followed by a gentle hump the other way in the chest region, called the thoracic kyphosis, and another arch in the neck called the cervical lordosis. The lumbar lordosis lessens with sitting when the pelvis tips backwards on the sitting bones (the ischial tuberosities) and increases with standing.

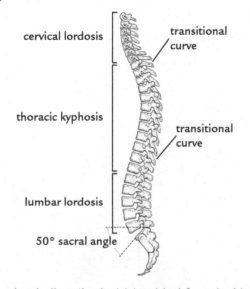

cervical lordosis

transitional curve

thoracic kyphosis

transitional curve

lumbar lordosis

50° sacral angle

Figure 1.2 A lumbar hollow (lordosis) is critical for spinal health; it allows the spine to sink and spring to absorb impact while walking and also stacks the spine in equilibrium during sitting.

Perfect lumbar alignment achieves two important ends: it ensures the correct distribution of bodyweight through the front and back compartment of each lumbar segment, and allows your lower back to bow forward slightly to absorb impact during walking. As you might imagine, the right amount of lumbar lordosis is an important factor in avoiding back pain.

The following discussion highlights the anatomy which allows the spine to move in its free-flowing way—guiding it and controlling it so it doesn't go too far.

The lumbar vertebrae

The vertebrae are the individual building blocks of the spine. Each has a front and back compartment. The front compartment consists of the circular vertebral body, shaped like a cotton reel, which is specifically designed to stack easily and bear weight. The back compartment protects the spinal cord and hooks the spinal segments together so they stay in place.

Five lumbar vertebrae make up the low back. At the base of the spine the bottom vertebra (L5) sits on the sacrum and the junction between the two is called the lumbo-sacral or L5-S1 joint. As the most compressed level in the spine it is the most problematic. A high percentage of back trouble is caused by dysfunction of the front or back compartment (sometimes both) at this level.

The back compartment is a ring of bone, which barely takes weight, extending backwards from the vertebral body. In standing it bears approximately 16 per cent of bodyweight, but less if the spine is more humped in the sitting position where the facets are less engaged. With severe disc narrowing—the primary form of breakdown of the spine— the facets may be forced to take much more weight (up to 70 per cent), which is tremendously destructive.

Each back compartment has small projections of bone sprigging from the outside corners: two wings out either side, called the transverse processes and a fin projecting out the back called the spinous process (these are the spinal knobs you can see through the skin running down the back). All these bony bars serve as levers for the attachment of muscles which make the vertebrae move.

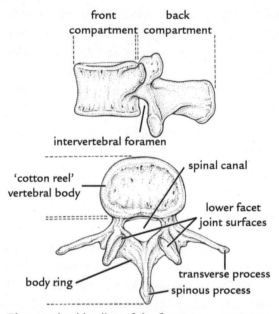

Figure 1.3 The vertebral bodies of the front compartment are designed to stack easily and bear weight, while the bony ring of the back compartment protects the cord and notches the segments together at the facet joints.

The 'cotton reels' superimposed on one another at the disc-vertebra union make up the beautifully bendable neurocentral core, and the junctions between are often called the interbody joints. The bony inter-notching either side at the back makes a chain of mobile juicy apophyseal or facet joints, running down the entire length of the spine. Together, the two different types of joints of the front and back compartment make up the total 'motion segment' at each spinal level.

All the muscles working the segments exert a downward pull as they bring about movement. If you bear in mind how much time we spend upright, fighting the weighing-down effects of gravity, you can see there are two factors at the start—our weight plus the muscular strings working the vertebrae—contributing to spinal compression. But there is also a third: sitting. Long periods of heavy sitting slumped in a 'C'-shaped posture loads up the discs and is particularly compressive; even more so as the facet joints at the back disengage and the belly lets go at the front. This is noteworthy, because I believe lumbar compression is the background cause of most low-back problems.

The vertebral bodies are prevented from grinding on one other by the intervertebral discs. These are high-pressure fibrous sacks containing an unsquashable sphere of fluid in the centre, called the nucleus. The 24 bony segments interspersed with discs makes the spine into a dancing resilient column, readily able to carry load and absorb extraneous forces from all directions.

The actual shape of the vertebral bodies helps spread the load. They have a narrow waist which flares out to a broad weight-bearing upper and lower surface. Unlike the other lumbar vertebrae, L5 is thinner at the back, which helps to form the lumbar lordosis. Its disc is also slightly wedge-shaped although it is still the fattest disc in the spine, helping it to bear the load of the rest of the body towering above.

Each 'cotton reel' is made up of a layer of hard cortical bone on the outside and honeycomb bone (or cancellous bone) on the inside. This is sometimes called the 'spongiosa' because it resembles a sponge and stores a rich supply of blood. The presence of the blood inside the bones is an ingenious way of dispersing forces through the bone.

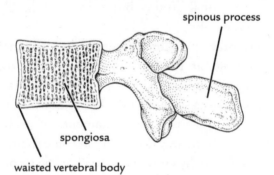

spinous process

spongiosa

waisted vertebral body

Figure 1.4 The honeycomb cancellous bone (spongiosa) is really a three-dimensional internal scaffolding which stops the vertebral body crumbling under load. The blood reservoirs in the vertebrae also help absorb shock.

Apart from being a handy reservoir, the fluid inside helps absorb the impact of shock passing through the vertebrae. These box-like bodies, literally bursting with blood, transmit the forces of compression in all directions throughout the fluid, thereby dissipating the direct downward pressure. As well as reducing strain, this functions as a useful engine for shunting nutrient-bearing fluids into the disc, which does not have its own blood supply.

The line of demarcation between the vertebra and the upper and lower surfaces of the disc is called the vertebral endplate. It is a thin cartilaginous interface about 2 mm thick and although each one is cushioned by the disc in between, it is still the weakest part of the spine. With rigorous impact, each endplate can seem like a semi-destructible membrane caught between two thundering fluid-transmitted systems: the vertebral body on one side and the disc on the other. Sometimes, impact through the spine can blow a tiny vent in an endplate, like blowing a hole through hide stretched over a drum.

The honeycomb bone inside the vertebra is actually a gridwork of tiny struts and spars, like internal scaffolding. Its three-dimensional structure prevents the roof of the vertebra caving in and the walls collapsing inwards like a cardboard box being flattened. It is a brilliant way of making the bones strong yet light. If the vertebrae were solid it would be much harder for our spines to operate. Not only would the bone tend to cleave off in chunks when subjected to compression and torsional strains but we would hardly be able to move for our own weight.

When the vertebrae are superimposed on one another, the consecutive bony rings at the back make a hollow tube inside the spine called the spinal canal. The canal houses the fragile spinal cord of the central nervous system which hangs down from the base of the brain like a long plait of hair. Filaments of nervous tissue branch off either side all the way down and become the spinal nerve roots. The cord itself actually ends at the level of the second lumbar vertebra. The roots then continue on inside the spine, hanging down like strands of a horse's tail (hence the name 'cauda equina') until they make their exit either side through their designated inter-segmental level.

Whereas the role of the front compartment is fairly straightforward as a weight-bearing strut, the workings of the back compartment are more complex. Apart from acting as casing to protect the spinal cord it has two other important functions: to guide the movement of the vertebrae—favouring some and keeping other more troublesome ones to a minimum—and helping to lock the vertebrae together to stop them slipping off one another.

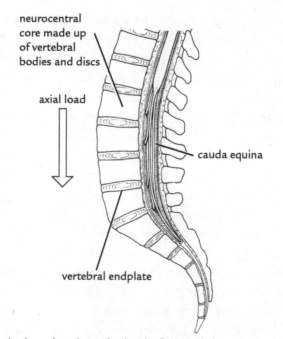

neurocentral
core made up
of vertebral
bodies and discs

axial load

cauda equina

vertebral endplate

Figure 1.5 The spinal cord ends at the level of L2 and the nerves continue on down inside the spine as the cauda equina. Each set of nerve roots exits the spine at its designated level.

The spinal ligaments

The spinal ligaments are a very important backup system to keep the spinal segments together. Between the bony locking mechanism and the muscular system they guide and restrain the movements of the vertebrae. The most important group is 'the posterior ligamentous system' made up of the capsular ligaments, the ligamentum flavum, the interspinous ligament, the supraspinous ligament, and the posterior longitudinal ligament. They make a dense festoonery of fibrous reinforcement connecting the bony parts of the spine at the back and the whole system comes into its own when the spine bends and lifts.

The capsular ligaments are really the facet joint capsules and they, together with the discs themselves, do the bulk of the ligamentous restraint in holding the segments secure when the spine bends. Researchers have removed the discs from cadavers in a laboratory and shown that half the body's weight can be suspended by facet capsules alone.

The ligamentum flavum is a thick short ligament covering the front of the facet joints at each segmental level. Each one has a smooth surface to help create a safe, comfortable lining for the back of the spinal canal with the delicate spinal cord inside.

In its healthy state, the ligamentum flavum is unusual because unlike other ligaments it has more muscle tissue (elastin 80 per cent) than fibrous (collagen 20 per cent), making it in reality a 'muscular' ligament. The purpose of this may be for ligamentum flavum to maximally shrink itself, thus avoiding physical runkling into the highly sensitive spinal cord. In the degenerative process, it is not uncommon for ligamentum flavum to calcify and become so bulky that it affects the very blood supply to the spinal cord itself. This is part of the process called vertebral stenosis.

While the ligamentum flavum covers the front of the facet joints, the multifidus muscle (about which you will read a great deal in this book) covers the back. With both muscles right at the nexus of forward movement of a segment, they are intimately involved in controlling it. In particular, they guard the laxity of the facet joints— which are the most likely part of the spine to come undone. They stand like muscular sentries front and back, controlling how much they pay out to let facets open for the spine to bend.

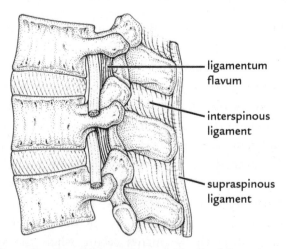

Figure 1.6 The collection of ligaments at the back of the spine constitutes the 'posterior ligamentous lock' which generates greatest holding tension when the back is rounded. Easy to see why we should lift with a humped lower back. (Note: Capular ligaments not shown.)

Both 'muscles' do an equally important job during bending by pressurising the disc. Through the facets they generate a tension at the back of the interspace which primes or pre-tenses the disc and stops unwanted wobble of the segment as the spine leaves the safety of vertical. It is a vitally important role and you will see in Chapter 4 how sometimes this mechanism falters. If the spine goes to bend before having worked up sufficient holding power in the multifidus muscle—usually due to bending with a weak tummy, or with the back arched instead of humped—there can be a fluke mishap where the upper vertebra slips imperceptibly at one of its facets.

The interspinous ligament is situated between the tails of the vertebrae with its fibres aligned in such a way as to resist opening of the backs of the vertebrae. The supraspinous ligament links the tips of the spinous processes (the tails), and also helps them to resist splaying open. There is no supraspinous ligament between L5-S1 interspace, presumably because this is taken care of by the massively strong ilio-lumbar ligament.

At the 'critical point' in bending, restraint passes from the spinal muscles to the posterior ligamentous system. Muscle activity ceases at full bend when you literally 'hang on your straps' with the back held stable by the passive tension of the ligaments. In the controversial area of correct lifting techniques it should be obvious that there are sound biomechanical reasons for slightly humping (rather than arching) the lower back to lift. This simple measure brings the critical point forward and makes the spine more stable earlier in range. (Though I fly in the face of convention here, I welcome—with some temerity—the debate opening up and ultimately benefiting our instructions to industry on manual handling.)

The ilio-lumbar ligament provides the main shoring for the base of the spine. It is a broad star-shaped band of fibrous thickening which passes from the inside bowl either side of the pelvis and converges upwards on the lowest vertebra, attaching itself strongly through the two huge transverse processes which curve downwards to meet it like tusks.

Incidentally, to provide extra strength in lashing the base of the spine to the sacral table, the transverse processes of L5 are built in a pyramidal shape with a broader base to receive the two strong ropes of the ilio-lumbar ligament. Handy as this may be in providing a better base for attachment, it does somewhat occlude the diameter of the intervertebral foraminae. These are the two small holes below the transverse processes at every spinal level through which the nerve roots issue. If you bear in mind that the L5 root is also the thickest, you can see at the start why the lumbo-sacral level is so prone to being inflamed by pathological processes affecting either the front or back compartment (or both).

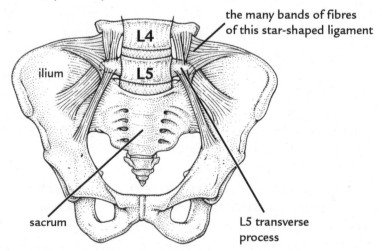

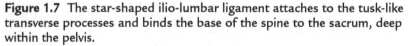

Figure 1.7 The star-shaped ilio-lumbar ligament attaches to the tusk-like transverse processes and binds the base of the spine to the sacrum, deep within the pelvis.

The long tube of 'cotton reels' making up the neurocentral core is reinforced front and back by two strap-like ligaments called the anterior and posterior longitudinal ligaments. The anterior longitudinal ligament is the strongest ligament in the spine and by interlinking the front of the vertebrae it stops your spine bending back too far. It also prevents the lower back sinking too deeply into an arch (or lordosis) as the spine takes weight.

The posterior longitudinal ligament runs down the back of the cotton reels, spreading out over the back of each disc in a cross-hatching of fibres to reinforce this part of the disc wall. More than the other ligaments, it has a highly developed nerve supply and is extremely sensitive to stretch. Significantly, in the case of a prolapsed intervertebral disc when escaping nuclear material squeezes out—

between chinks in the disc wall (see Chapter 5), the cause of irritation can be twofold: stretching both the outer layers of disc wall and the ultra-sensitive posterior longitudinal ligament lying on top of it.

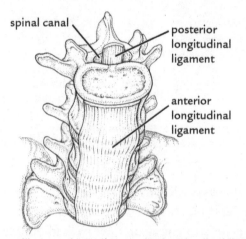

spinal canal

posterior
longitudinal
ligament

anterior
longitudinal
ligament

Figure 1.8 The strap-like anterior and posterior longitudinal ligaments encase the front and back of the 'cotton reels' like a ligamentous strait-jacket.

At its simplest, the spine gets most of its movement from the cotton reels sitting on their discal pillows and careening about in all directions. The role of the back compartment of each motion segment is to control that movement.

The intervertebral discs

The intervertebral discs are resilient, water-filled pillows between the vertebrae. They have scant nerve supply and no blood supply; indeed discs are the largest avascular structures in the body. This also means that other mechanisms must be utitilised to transport raw materials and waste products to keep the discs alive, which will be discussed below, but even at the best of times discs struggle to remain viable.

Discs are designed to give us flexibility at low load and stability at high load and it is significant that our vertical posture helps achieve this by enhancing the tensile strength of the disc walls. It adds to the pressurising of the fluid sacks and converts the spine into a whippy spring-loaded rod of many segments which can flip up straight again after bending.

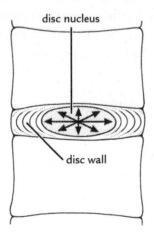

disc nucleus

disc wall

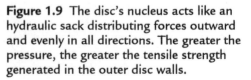

Figure 1.9 The disc's nucleus acts like an hydraulic sack distributing forces outward and evenly in all directions. The greater the pressure, the greater the tensile strength generated in the outer disc walls.

Without these tensile properties, the human back would not be the long slender thing it is. We would need a hugely muscular apparatus to haul us up straight again once we had doubled over. However, vertical posture does have its down side. It means the segments at the bottom of the stack are always more loaded by the weight of the rest of the body towering above.

High intradiscal hydrostatic pressure of the discs thrusts the vertebrae apart while at the same time as it helps glue them together. Each has a vigorous incompressibility, like standing on a breadboard balanced on a beachball. They give the spine a quivering up-thrusting romp, while making it secure enough to bend without getting stuck, flopping over like a broken reed.

The squirting liquid nucleus of each disc is contained by a tough, circular containing wall called the anulus fibrosis. The anulus is like an onion skin, made up of approximately 12–15 thin fibrous layers called lamellae. The fibres of each successive lamella run in diagonally opposing directions at an angle of 65 degrees to the horizontal. This provides maximum strength, yet freedom for the wall to pull up, like a lattice stretching, as the spine bends. This fibre angulation makes the disc harder to stretch than if they ran transversely, but easier than if the fibres were aligned vertically to the direction of movement.

The lamellae at the back of the disc are thinner and bunched more closely together, making it possible for the wall to stretch vertically by

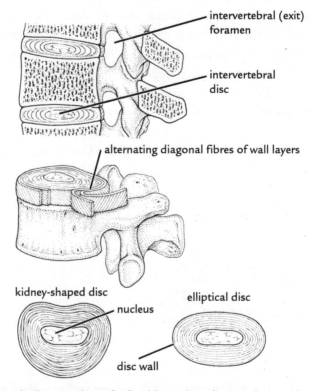

intervertebral (exit)
foramen

intervertebral
disc

alternating diagonal fibres of wall layers

kidney-shaped disc

nucleus

elliptical disc

disc wall

Figure 1.10 Each disc consists of a liquid pearl nucleus and a tough multi-layered wall which helps keep the nucleus contained under pressure.

50 per cent. This provides freedom for the interspaces to gap open at the back so the spine can bend forward freely. It also means it is weaker, introducing a precarious trade-off between freedom to bend and the possibility that over-bending can break down the wall.

Often the L5 discs are kidney shaped, which exposes a longer flank to increasing the holding power of the back wall. However, kidney-shaped discs have the disadvantage of runkling more in the acute back corners when torsional strains are applied to the disc. You will see later how heavy-duty lifting and twisting can make the disc wall perish at these points.

The roles of the inner and outer disc walls are very different. The middle/inner anulus creates a super-strong capsule which helps hold the watery nucleus contained under pressure. The lamellae here continue in a circular fashion, around both sides of the disc and through the endplates above and below, thus trapping a buoyant hydrostatic pressure within the disc to force the segments apart. This part of the disc bears, indeed rebuffs, compressive axial load. Conversely, the outer annulus works like a tensile ligamentous 'skin' that holds adjacent spinal segments together. Just like any other ligament in the body, this part of the anulus restrains the bones from moving apart and has no role to play in shouldering load. Indeed a low intradiscal pressure can cause the outer disc wall to buckle outwards under load (or when compressed by muscle spasm) and this is often wrongly construed as a prolapsed disc.

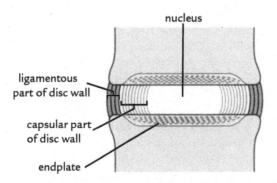

ligamentous
part of disc wall

capsular part
of disc wall

endplate

nucleus

Figure 1.11 The middle-inner part of the disc wall acts like a capsule to bear load whereas the outer part holds the segments together and is more like a tensile skin. Like any ligament, this 'ligamentous' part of the disc can scar and adaptively shorten, which makes that segment become a stiff link in the spine.
(Illustration acknowledgement: N. Bogduk, 'Clinical Anatomy of the Lumbar Spine')

Also like any other ligament, this part of the disc has a nerve supply and readily registers pain. The outer anulus complains if wrenched or overstretched by trauma (recall the pain of a twisted ankle) and also becomes painful if it adaptively shortens and cannot 'give' with movement. This is what happens when a disc dries and loses height and this part of the back then becomes painful to bend. As you will read in the next chapter, discomfort is invoked as the spinal segments attempt to pull apart and eventually inflammation sets up at that segmental level, which I believe is the nub of simple back pain.

In the healthy state however, both parts of the disc wall work well to complement each other. A high intradiscal hydrostatic pressure 'inflates' the central capsule and pre-tensions the outer ligamentous wall, thus making it hold more securely with the right amount of hold-and-give. Just as inflating an inner tube of a tyre gives the outer wall suitable tensile strength, the spine can bend and sway freely with the internal pressure of the disc matched by the tension of the restraining disc walls, and everything moves under control. If a lack of internal disc pressure fails to invoke sufficient holding wall tension of one of the discs, that segment can become loose and shear forward as the spine bends. This is the main cause of an unstable spinal segment.

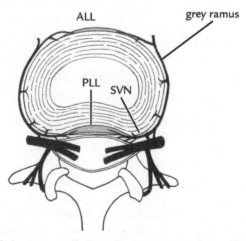

Figure 1.12 Only the outer tensile ligamentous 'skin' of the disc wall is pain sensitive.
(Illustration acknowledgement: N. Bogduk, 'Clinical Anatomy of the Lumbar Spine')

Proteoglycans is the magical x-factor that attracts and holds fluid in the discs. Its unique molecular make-up exerts a powerful osmotic pull on water that counters the effects of gravity bearing down and squeezing the discs dry. Healthy nuclear jelly taken from a disc and set in a saucer of normal saline solution swells by 300 per cent and this force within the discs becomes more potent through the day as they lose fluid and the proteoglycans concentration rises. Even so, all discs lose about 20 per cent of their fluid each day, which they recoup when we lie flat to sleep overnight. This steady seepage of discal fluid by day, with fresh quantities absorbed again overnight, is the main way discs nourish themselves; this stately exchange is only possible because the metabolic rate of discs is so slow. Healthy discs have a high concentration of proteoglycans whereas degenerated discs do not. This means that degenerated discs are drier, lose their fluid more rapidly during upright hours, and are slower to regain it when compression comes off. Low proteoglycans discs become thinner, poorer spacers, poorer shock absorbers and weaker spinal connectors.

Very importantly from a therapeutic point of view, synthesis of proteoglycans is stimulated by *pressure changes* through the discs. Normal on–off gravitational forces sustained by the spine as it bends, twists and lifts everyday things enhance both the fluid content of discs and the movement of fluid through them, as well as stimulating disc cell metabolism in general. This simple physiological truism (called 'mechanobiology') gives discs their best chance to regenerate or *heal* and its significance is essential to the understanding and implementation of effective 'spinal therapy'. It underlines the main focus of self-help treatment and the various means of achieving it are explained in each of the following sections.

The nutrition of the discs

The tenuous viability of discs, even in their healthy state, means that nutrition is critically important for coping with the ongoing incidental trauma of everyday life. Poor disc nutrition has long been held as the chief cause of disc degeneration, so it follows that physical therapy is about improving disc nutrition—specifically by enhancing the disc's ability to hold and circulate water.

There are two main engines of fluid exchange whereby nutrients are absorbed and waste products expelled from discs. The first is the 'diffusion' mechanism described previously, which is transacted as a diurnal exchange over 24 hours; stale fluid slowly excretes through the day and fresh quantities are imbibed overnight. During the day, fluid is pressed out through the disc walls and vertebral endplates, and as they become drier there is a vertical settling of the column as the discs flatten and the spine as a whole loses height (we are all approximately 2 cm shorter by the time we go to bed). This engine is mainly effective in transporting smaller molecules involved in disc metabolism, such as glucose and oxygen.

The second engine is more active through the daylight hours.

The 'convection' method relies on pressure changes induced by grand-scale spinal movement. It creates a rhythmic suction and compression of the discs which shunts small quantities of fluids in and out from the rich capillary beds in the neighbouring vertebral bodies. This method transports larger molecules which have poor diffusivity, the most important of which are cytokines involved in the manufacture of proteoglycans.

This secondary 'pump imbibition' is an invaluable backup mechanism for nourishing discs, especially in circumstances where they suffer undue compression. For people who sit for many hours, particularly sedentary workers or long-distance drivers (who, incidentally, have back problems at four times the national average), squatting or curl down (touching toes and unfurling back up to vertical) exercises are particularly important ways of introducing pressure changes through the low back. The lumbar discs are subject to greater pressure in the early part of the bend but as the spine gets to the bottom and hangs the effect is one of relative suction. Activities of daily life are just as useful. No matter how humble the task, dynamic loading and releasing during spinal movement sucks and pumps the discs through the day, recouping small amounts of fluid and thereby retarding the inexorable settling of the spine.

Whole-scale spinal movement and its shunting effect on discal fluids become even more necessary in three important circumstances:

- when ageing causes occlusion of the small holes in the endplates through which fluids are exchanged from within the vertebral body
- when ageing causes fibrosis of the disc walls making them less permeable to fluid traffic
- when disc degeneration lowers the concentration of proteoglycans which lessens the disc's water-binding capacity.

Flamboyant spinal movement (with as little jarring impact as possible) keeps the spine young and 'gives the discs a drink'. Sadly, one of the oldest directives in orthopaedic medicine and the management of bad backs has been to limit movement—and this has directly hampered disc nutrition and the regeneration of spinal health.

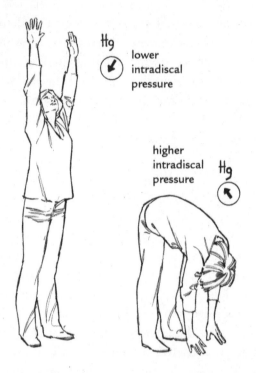

lower
intradiscal
pressure

higher
intradiscal
pressure

Figure 1.13 Pressure changes induced by everyday bend-and-stretch activities help the discs suck and blow to feed themselves. They also stimulate the metabolic activity of discs.

The vertebral movements

The movements of the vertebrae in the spine are a combination of gliding, tipping and twisting, although each one individually contributes only a small degree. Superimposed one on one however, the net result is the grandiose wide-ranging mobility of the spine, which is so well known to us. From our towering height we can arch backwards under a limbo bar and bend over to cut our toenails. Well, some of us can.

Of the vertebral movements, gliding is the least generous. The upper vertebra slides transversely forwards, backwards and from side to side on the vertebra below but the actual distances travelled are minute. Glide is more of a background movement which better positions the vertebra for action; it sets the stage and puts the

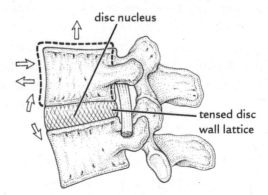

disc nucleus

tensed disc
wall lattice

Figure 1.14 The vertebral bodies roll around on their disc while the facets at the back act like guide rails to keep the movement in check.

vertebra at its optimum starting point for the more adventurous tipping and twisting activity to follow.

As we bend to touch our toes, for example, each vertebra moves forward incrementally on the one below, bringing the upper one to its perfect starting point for tipping. The element of glide contributes the typical flowing quality to the way all living creatures move (a cheetah on the run has it in spades). Without it, all living actions are much more clipped and jerky and not nearly as streamlined and expansive. An elderly lady tottering along the footpath has very little glide in her joints.

The right amount of vertebral glide is important; it is what healthy backs have. Too little or too much glide leads to trouble. If a segment has too little it will be stiff. And significantly, when the degenerative process sets in, it is the first movement to go. Although you cannot necessarily see it—first you only feel it—this lack of background movement makes your spine feel tighter; more restricted and laboured in everything it does. In short it makes you feel rigid and older than you are.

At the other extreme, a vertebral segment is unstable if it has too much glide. It comes to light when the spine bends over and the top vertebra slips forward on the lower one. This is known as segmental instability, and it actually stems from stiffness of a segment. What distinguishes the former from the latter is the total degeneration of the motion segment. It is the stuff of this book and discussed step by step throughout.

With normal segmental mobility, all the vertebrae roll around on their liquid ball-bearing discs whose fibrous walls keep everything in place. As the upper vertebra moves off-centre, tension takes up in the wall mesh and brakes the action. And as soon as bend is incorporated into the action, another brake comes on from the squirting pressure of the nucleus.

With bending forward, which is our most-often repeated action, a healthy nucleus squidges under pressure towards the back of the disc. The migration of this buoyant bubble of fluid thrusts up the back of the vertebra above, thus 'bending' the segment in a tipping action. The opening of the back of the interspace also increases the tension-hold of the mesh of the disc wall and invokes powerful restraint from the capsular ligaments of the facet joints.

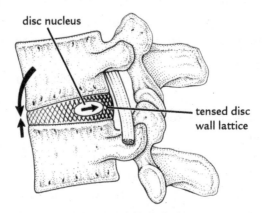

Figure 1.15 The bending action invokes its own brake when the backward migration of the nucleus pre-tenses the back wall of the disc. This makes it harder for the two vertebrae to pull apart.

Thus, a well-hydrated disc provides a high intradiscal pressure which adds dynamic stiffening to both the disc wall and the posterior ligamentous system. It also provides good cushioning and clearance between its two neighbouring vertebrae. So, you can see how a well-hydrated disc improves the restraint functioning of both the front and back compartments and in fact creates 'stability' of a working segment. On the other hand, if a nucleus lacks sufficient water, the disc narrows and the segments will settle together, bone on bone. It also fails to provide sufficient oomph to spring-tip the upper segment as the spine bends and the segment will shear forward instead of tipping. As you will read in Chapter 6, this is usually how a degenerating disc becomes over-mobile in the spinal column and eventually becomes unstable. Segmental instability is the end stage of segmental breakdown.

The facet joints

Each motion segment has two facet joints forming the back compartment. These are the junctions formed where the vertebrae notch together at the back of the spine. They flank the back corners of each disc, across the gulf of the vertebral canal. The bony notching of the facets provides a primitive interlinking chain down either side of the spine which slots the spinal segments together. If they were not there the vertebrae could roll around on their discs and the neuro-central core could tie itself in knots like a cartoon character of an India-rubber man. The lumbar facet joints do not bear a lot of weight unless the disc is thin or the lumbar lordosis extreme, but they suffer constant wear and tear in controlling the movement of their vertebrae.

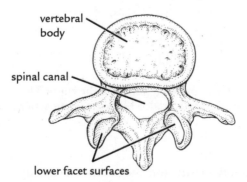

vertebral body

spinal canal

lower facet surfaces

Figure 1.16 The lower facet surfaces are concave to receive the convex surfaces of the upper facets notching in. The paired facets flank the back of the spinal canal at each segmental level.

Neighbouring vertebrae contribute two opposing surfaces to make a pair of facet joints. Two notches of bone project up from the lower vertebral ring (superior articular pillars) interfacing with two projecting down from the upper vertebral ring (inferior articular pillars). The smooth-interfacing congruency of these four abutting pillars forms the two facet joints. They fit together snugly like two cupped palms, the valleys of one side matching the hills on the other. Each is covered by a buffer of hyaline cartilage with its ultra-smooth semi-compliance and rich mother-of-pearl sheen.

The convexity of the upper facet surfaces and the concavity of the lower ones allow the upper vertebrae to slot into place. The definite front–back alignment of the lumbar facets facilitates forward bending but restricts all other movements of the low back. It means the vertebrae only move forward and back, like the wheels of a train moving down the track, never twisting left or right (although they can side-bend a little). It lets us lower the upper body forward, like a stooping mechanical crane, putting our hands and face at the right height to be useful.

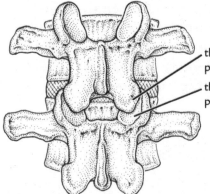

the inferior articular
pillar of the vertebra above

the superior articular
pillar of the lower vertebra

Figure 1.17 The front–back alignment of the cupped lumbar facets facilitates forward bending. It also blocks twisting actions which weaken the discs.

There is good reason for the facets restricting lumbar twisting movement: it keeps things from wearing out. The twisting action in the low back is especially harmful to the discs. It challenges the inherent weakness of their walls, especially if there is lifting as well.

With only every alternate layer of the 'onion-skin' disc wall offering restraint (while the fibres of the other half go on the slack offering no help), repeated twisting can be destructive.

The most capable and resilient joints in the body are synovial joints, and all of them—whether of the fingers, knees or facet joints—share common properties. They have a smooth cartilage buffering on opposing interfaces to minimise friction and they are knitted together by a strong fibrous sleeve or capsule, the internal lining of which secretes a lubricating synovial fluid.

Cartilage allows opposing facet surfaces to skid over one another with the yielding consistency of dense plastic, so that it deforms imperceptibly whenever the bones make contact. The direct contact also squeezes fluid out, but when the pressure releases and the cartilage un-dints, it sucks water back in. In this way the bloodless cartilage keeps itself healthy by creating a 'circulation' to pull in nutrients and expel waste products.

Synovial fluid—which has astonishing qualities of lightness and slipperiness—lubricates the cartilage interfaces and flushes them clean, similar to the way tears water the eyes. The strong joint capsule keeps the fluid contained under pressure that springs the joint surfaces apart and softens the impact of bone on bone. It also means the joint operates on a cushion of fluid (in an hydraulic sack) which streamlines movement and takes out the jerkiness.

Synovial fluid also cleanses the joint space by clearing away cartilage particles eroded off the main bed during activity. The synovial membrane liberates large cartilage-eating cells (macrophages) into the tide of floating debris. These cells surround each particle, like an amoeba trapping its food, and dissolve it. It is essential cleaning-up work. Without it the joints would silt up with cartilaginous grit acting like a pot-scourer, grinding away the joint surfaces until nothing was left.

The facet joint capsules are stronger than most synovial joint capsules and for this reason they usually go by the name of the 'capsular ligaments'. Almost equally with the discs, the capsular ligaments share the role of restraining the lumbar segments as the spine tips forward to bend, the disc holding the front compartments together and the capsular ligaments the back. In addition to their

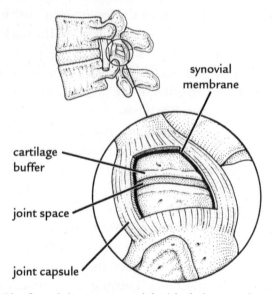

synovial
membrane

cartilage
buffer

joint space

joint capsule

Figure 1.18 The facet joints are synovial with their opposing cartilage-covered surfaces fitting snugly together like two pressed-together palms. The facet capsule holds the joints in place.

important ligamentous role, the facet capsules are fully innervated (well supplied with nerves) and have a prolific blood supply. Quite unlike the discs, the facets are wired for pain and have the propensity for florid inflammatory reaction in response to physical assault. For this reason, the facet joints are a common source of low-back pain, especially if the disc has started to degenerate and, in losing pressure, transferred a greater restraint role to the facet capsules.

The bending human spine

Elegant and strong as the human back is, the job of bending and straightening is a tall order. The trunk and spinal muscles which actively control the movement are discussed further on. However, several other anatomical features help make bending possible, by working as a physical brake to control the free fall of the spine when it tips forward.

The first of these we have discussed already: the strong fibrous wall of the disc which holds the cotton reels together. This contributes about 29 per cent to the control of the segments going forward.

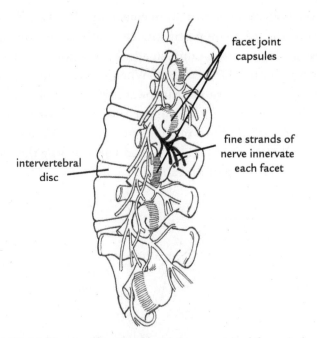

facet joint
capsules

fine strands of
nerve innervate
each facet

intervertebral
disc

Figure 1.19 Unlike the discs the facet joints are wired for pain by the medial branch of the lumbar dorsal ramus.
(Illustration acknowledgement: N. Bogduk, 'Clinical Anatomy of the Lumbar Spine')

As the segments glide forward the stiff fibrous mesh of the wall retards the initial movement. When the spinal segments then tip forward and the back of the interspaces opens up, the same diagonal mesh pulls up, like stretching up a garden lattice.

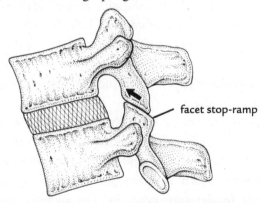

facet stop-ramp

Figure 1.20 As the spine bends, the facet surface of the upper vertebra slides up the stop-ramp of the one below, increasing the tension between the two segments and bringing movement to a halt.

However, even more important in controlling bend are the various structures of the back compartment. The bony facets contribute in two ways: a sloping stop-ramp, made by the joint surfaces and extremely tough capsular ligaments. When viewed from the side, the lower facet surfaces taper upwards towards the front of the spine. As the spine bends this means the upper vertebrae must travel uphill as they go forward. (They work like emergency stop-ramps beside steep downhill sections of highways, gradually bringing the vehicle to a halt as it nears the top of the ramp.)

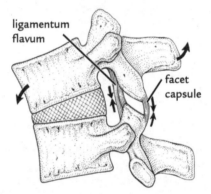

ligamentum flavum

facet capsule

Figure 1.21 The capsular tension of both the facet capsules (known as the 'capsular ligaments') and ligamentum flavum holds the facet surfaces pressed together and restrain the back of the spine opening as we bend.

In the case of the back, the escalating tension of the soft tissues gradually brings the upper vertebra to a halt, by which time the facet interfaces are firmly locked against each other and the ligamentum flavum and the facet capsule are tense at full stretch. It is a marvellously ingenious system with both bone and soft tissue complementing the workings of each other.

As we go further into a bend the upper vertebra then tips bodily forward by pivoting on the front edge, as the tail of the vertebra attempts to lift away. This second part of the movement is restrained mainly by the facet joint capsules. They contribute a powerful 39 per cent towards moderating bending. The ligamentum flavum contributes an initial 13 per cent. All up, the facets contribute 52 per cent of the ligamentous restraint on forward bending.

It is very significant that none of these facet ligaments exerts control over the forward *gliding* action of the segments, only forward tipping. Control of forward gliding is important because it is this action carried out to excess (when it is called 'forward shear') which constitutes the unstable element of a segment's movement. All segments must avoid shear, because it is potentially so devastating. And it is failure to control the forward tipping which allows too much shear.

The process works like this: as bending starts, the initial forward glide is only about 2 mm before the facets engage to stop it. As the bend continues, the facets disengage with the tipping forward action which makes the tail of the vertebra lift up and away, leaving a gap between the two facet surfaces. However, once the tail of the vertebra is away, the whole vertebra can glide forward more, until the bony block engages once again. In this way, more tip allows more glide. Incidentally, the tipping action is what both the facet capsules and the ligamentum flavum are designed to resist, while multifidus, the deepest intrinsic muscle of the spine, pays out to control it actively.

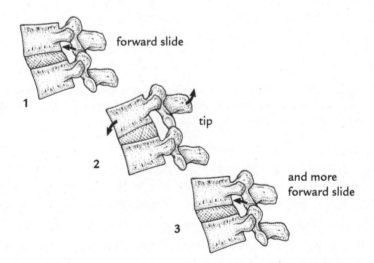

forward slide

tip

and more forward slide

Figure 1.22 The segments have scant ligamentous shoring to control forward shear but plenty to restrain forward tip. Controlling shear of the segments is thus controlled by restraining tip.

The spinal nerves

In the lumbar area, the nerve roots emerge from the spine under their corresponding vertebra. Thus the left and right L1 nerve roots come out under the first lumbar vertebra at the L1-2 interspace and so on. The L5 nerve roots come out at the lumbo-sacral junction. The spinal nerves carry messages to the muscles to make the legs work and also carry sensory messages back inside, relaying information from the outside world back to the brain.

As the nerve roots leave the spine they travel out through small canals (intervertebral foraminae) bordered on one side by the facet joint and on the other by the disc. It is less than ideal to have these fragile strands of nerve making their exit right through the machinery of a complex human hinge. It means they run the gauntlet between the very two structures most likely to cause trouble in the spine.

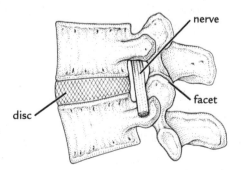

Figure 1.23 The spinal nerve root exiting the spine passes between the disc sucking and blowing on one side and the facet capsule pulling and puckering on the other. This is a precarious arrangement.

As each nerve root goes about its business, it has the intervertebral disc sucking and billowing on one side and the facet joint with its baggy capsule tugging and puckering on the other. The nerve is largely protected from these goings-on by a protective sleeve or nerve sheath which extends just beyond the spine, like the cuff of a shirt poking out from a coat.

As you will read in the next chapter, pathological change of either the disc or facet joint can cause thickening, hardening and swelling.

By direct contact, the inflammation can spread to the nerve root, simply because it is so close. Inflammation of the nerve causes severe pain down the leg, known as sciatica.

The muscles which work the spine

As superbly designed as the spine is, it amounts to naught without the dynamic contribution of the muscles. In the way a puppet is a flummoxed pile of sticks on the floor without its working strings, the human spine and its segments are an inert, toppling pole without its muscles.

The muscles of the human body work just like the strings of a puppet. They pull on levers and make the body move. They allow us to keep the thinking, top part of the body up there and active so we can operate effectively in the outside world. Without the dynamic support of the muscles the spine would fall over. More than you would ever imagine, the muscles play a dynamically synchronised role in keeping the skeleton upright and controllable. You only have to see unfortunate cases of poliomyelitis to understand this point.

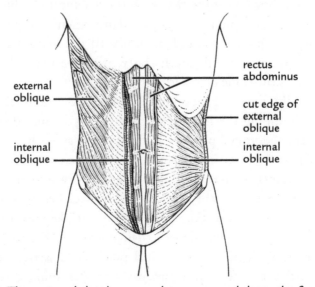

Figure 1.24 The rectus abdominus muscle runs up and down the front of the abdomen whereas the internal and external oblique muscles cross over each other diagonally making a wide 'X' and drawing in the sides of the waist.

With paralysis of the trunk muscles, the spine tumbles down around itself in slow motion, like a collapsing circular staircase, with the chest disappearing into the abdomen and into the pelvis. The tummy muscles play a critical role in keeping the towering spine up.

Tummy muscles also play a critical role in letting the spine bend. Although there is some confusion about the mechanism by which this happens, practising therapists the world over recognise the importance of a strong tummy in treating back problems. Still today, researchers cannot agree about the best way to strengthen these muscles. I certainly have my ideas which you will read about later in the book.

The abdominal muscles spread vertically, horizontally and diagonally, wrapping the soft abdomen in sheets of contractile tissue. As they contract, their fibres shorten and bow the spine forward. Working statically they nip in the waist, creating the 'hourglass' figure and flattening the belly. This effect reduces the intra-abdominal space, which automatically raises the intra-abdominal pressure.

A strong tummy stabilises the spine from in front, in several ways. Firstly, a strong co-contraction with the back muscles lifts the spine vertically. This is similar to the upward movement of water in a plastic bottle if you grasp it around the waist. (If the bottle has no lid the contents spurt out the top.) The tension between the segments increases as the spine grows longer.

Secondly, a strong abdominal contraction pulls the belly in, slightly humping the low back and creating a contained pocket of high intra-abdominal pressure in front of the spine, like an air bag in a car. Apart from stiffening the spine, the back-pressure against the lumbar segments prevents them shearing forward on one another as the spine bends. As the bend of the low back becomes more accentuated and the tails of the vertebrae fan out like fish scales flaring, the posterior ligaments achieve their maximum tension and create a ligamentous lock. At the critical point, load is transferred from the muscles to the taut ligaments and the spine then becomes more stable.

As we bend, the abdominal muscles and their opposite number, the erector spinae muscles down the back of the spine, help in a

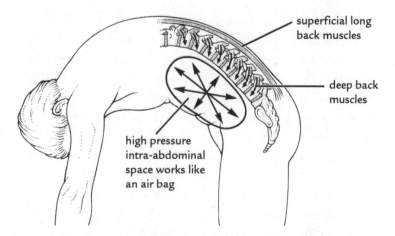

superficial long
back muscles

deep back
muscles

high pressure
intra-abdominal
space works like
an air bag

Figure 1.25 You must always 'unfurl' to straighten with both the long erector spinae muscles and the smaller segmental intrinsic muscles working optimally around the hoop of the spine. The tummy must be pulled in hard.

coordinated way to lower the spine over. From the front the tummy muscles provide the hydraulics for the fine precision. With fantastic, noiseless accuracy they allow us to adjust our height, while the long cables of erector spinae let the spine go, like a mechanical crane paying out. They particularly come into their own around the hoop of the spine when we are nearly at full bend.

Whether the bend is great or small, lowering the top-heavy column and then straightening it is a stupendously difficult task, especially when accompanied by lifting. Some mathematical calculations suggest the smallness of the long back muscles and the awkward angles make it impossible. Since we all know that our spines do bend and lift, and fairly effortlessly at that (though some better than others), we at least must concede the working spine is an awesomely effective thing.

With bending, the muscles of the spine and trunk work in two distinct ways: they centrally clench the segments together to keep them stable, and then they control the lowering over of the whole column. This vertical clenching of the spine is an important preparatory contribution to stability, while the bending itself is almost free fall. Rather than actually doing anything, the main job of the muscles is keeping control as the column goes over, so nothing comes undone.

Weakness in the central clamping-down action is one of the earliest things to go wrong with a back once it starts to break down. We do not know yet why this is so; whether it is a case of cause or effect. In other words, we cannot say whether the muscles reflexly inhibit once there is inflammation between segments, or whether the inflammation develops because there is weakness of the muscles keeping the segments together.

Coming up from the bend the whole process works in reverse. While the spine is still stooped, the ligamentous lock holds it stable and the unfurling starts by the pelvis rolling back. Then the static tensing of the abdomen thrusts us up from in front and the intrinsic muscles of the spine work at segmental level pulling each vertebra back. The long erector spinae muscles work around the hoop of the spine when we are in full bend, and then again like guy ropes once we are upright.

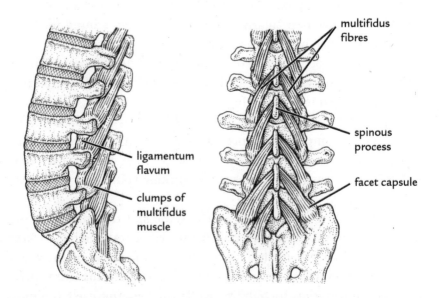

Figure 1.26 The deeper fibres of multifidus originate on the spinous process two vertebrae above and pass down and laterally to blend into the back of the facet capsules. Multifidus is the 'guardian' of facet opening and closing.

Multifidus is—especially in the early part of range—the most important intrinsic spinal muscle. It originates either side of the spine, on the bony ring of the back compartment, right over the top of the facet joints. It is the deepest segmental muscle and its fibres actually blend in with the back of the facet capsules (which is significant with facet trouble, see Chapters 3 and 4). The fibres then pass upwards and inwards to their vertebra where they attach themselves to the spinous process. With the spine upright the line of pull of multifidus fibres is at 90 degrees (or right angles) to the spinous process it is attached to. Thus it is optimally placed to control the segments at the very first glimmer of bending.

In straightening from the bent-over position, multifidus initially gets a bit of help from the muscular ligament, the ligamentum flavum, each working its own side of the joint. The smaller front muscle tries to close the joint by sliding the two joint surfaces together, while the larger multifidus closes the gap by pulling on the tail at the back of the spine.

Multifidus's main role during straightening is to pull down on the tail of the tipped-forward vertebra, like pulling down an overhead garage door. As we straighten, the spine unfurls step by step in a segmental fashion, from the hooped-over position to fully upright. Multifidus is vitally important in controlling the stability of the spinal segments, and it works hand in glove with the important stabiliser of the front compartment, transversus abdominus.

Multifidus does another unique job as it helps the spine straighten. It hoists the facet capsules out of the way, rather like a damsel gathering up her petticoats as she goes to climb the stairs. This prevents the tender capsular lining being nipped as the opposing surfaces close down when the spine straightens. As you will read in Chapter 4, sometimes the muscle's coordination is caught off-guard and the capsule gets painfully jammed in the works. It is thought that the muscular ligamentum flavum does a similar thing on the other side of the joint and prevents the baggy capsule getting nipped as the spine recovers from the arched (extended) position.

The other intrinsic muscles, iliocostalis and longissimus, control the forward shear of the vertebrae although they operate at a more difficult angle when the spine is more deeply bent forward. Their

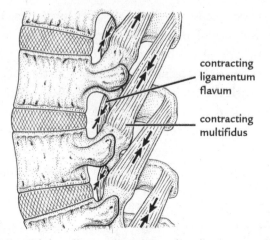

Figure 1.27 With bending and straightening, ligamentum flavum and multifidus work right at the centre of movement to make sure the facet capsule does not get nipped in the joint.

fibres pass in a direction closer to a back–front alignment. As the spine straightens they slide the vertebrae backwards in a reverse shearing action, rather like sliding drawers out of a chest. The deepest fibres of erector spinae muscles have a more transverse angulation than the superficial and they also contribute segmentally to controlling forward glide.

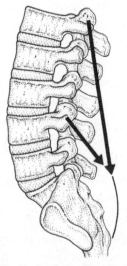

Figure 1.28 The direction of the fibres of iliocostalis and longissimus helps slide the vertebrae backwards as the spine straightens but their poor angle of pull makes their action weak.

The deepest tummy muscle, transversus abdominus, plays a unique role as it helps the spine straighten, because it works both as a tummy and back muscle at the same time. It originates from either side of the linear alba (the shallow groove running down the centre of the abdomen) and passes transversely around the waist below navel level, like a cummerbund. Either side at the back it joins a sheet of fibrous tissue, which performs a very important role, called the thoraco-lumbar fascia. The fibres of this fascial sheet make a diagonal lattice which attaches in different layers to both the transverse processes of the lumbar vertebrae.

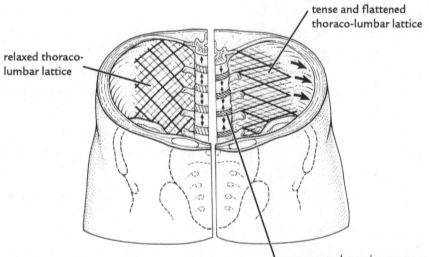

tense and flattened
thoraco-lumbar lattice

relaxed thoraco-
lumbar lattice

neurocentral core is compressed

Figure 1.29 Transversus abdominus constitutes the deepest layer of the tummy muscles. As a 'muscle-in-the-round' it performs two actions in one: drawing the belly in at the front and straightening the spine via tightening the thoraco-lumbar fascia at the back.

As the muscle contracts it creates a high pressure air bag in front of the spine (like any other tummy muscle). But it also tugs sideways on the thoraco-lumbar lattice. As the lattice pulls out laterally it becomes shallower in height which telescopes the cotton reels down on one another and tugs down on lateral spars of the vertebrae, thus clenching the spine.

As well as helping multifidus regulate forward bending, transversus abdominus performs the very important role of clamping of the spinal segments together in preparation for spinal activity. This ensures the spine doesn't jump out of joint as soon as other movements pull it about. This clamping works in all sorts of ways, from the subtle to the spectacular. For instance, it gathers up the spinal segments and allows us to turn over in bed at night without waking, but it also stiffens your spine automatically as you see the tennis ball coming at you over the net. It converts the lumbar segments into a stiffened pillar in anticipation for other things it has to do. With its tendency to weaken, no wonder these actions (to name but a few) are so painful with bad backs.

Recent research suggests that transversus abdominus is supported in keeping the segments stable by the ever-acting breathing muscle, the diaphragm. In an example of Nature's expediency, both take turn about in keeping the lateral tension on the thoraco-lumbar fascia. Since we breathe all the time I cannot imagine a better partner. What could be more appropriate than harnessing the breathing mechanism to assist another equally fundamental mechanism: keeping the central strut of the body intact.

It happens like this: the diaphragm attaches in part to the sides of the thoraco-lumbar fascia. It is a huge dome of contractile tissue which separates the thorax and abdomen, making both into water-tight compartments. We take a breath in when the diaphragm contracts and flattens and descends in the chest. This increases the volume in the chest cavity but lowers the pressure, causing air to flow in through the nose. The contracting diaphragm tugs laterally on the sides of the thoraco-lumbar fascia, making each in-breath telescope the lumbar segments. As we breathe in, the low back is kept secure as we go about our business.

When it is time to breathe out the diaphragm relaxes, raising the pressure inside the chest compared to the outside, making air flow out. At the same time, the diaphragm hands over the reins to transversus abdominus which is active during the breathing out phase, but which also exerts tension on the thoraco-lumbar fascia. During expiration it shrinks the girth which automatically raises the intra-abdominal pressure, helping to stabilise the spine.

Thus with each 'muscle' taking turns to keep the thoraco-lumbar fascia tense, we keep our backs stable just by breathing. This also explains why it so often hurts to cough with a bad back. The explosive exhalation happens with a strong involuntary contraction of the tummy which jerks the thoraco-lumbar fascia sideways. This bounces the vertebrae vertically and invariably elicits pain from the problem part.

The breathing control of the thoraco-lumbar fascia also explains why we automatically hold our breath when we lift. The held breath and the clenched tummy recruit both muscle systems simultaneously, raising the tone in the thoraco-lumbar fascia. The lumbar segments are held doubly secure.

This is exactly what weight-lifters do. As they bend to grasp the bar they take a sharp breath in and hold it, sucking their tummy in at the same time. Some professionals even wear a 'kidney belt' to reinforce the power of transversus abdominus, making it easier to generate the power to slot the lumbar vertebrae back on one another as the muscle pulls the spine straight.

2

The stiff spinal segment

This is the first stage in the breakdown of the spine, when one vertebra becomes stiffer than the rest, like a stiff link in a bicycle chain.

WHAT IS A STIFF SPINAL SEGMENT?

In the well-oiled spinal chain, a stiff spinal segment is a sluggish vertebra which participates less willingly than the others in overall spinal movement. More often than not the stiffer one causes no trouble; it just sits there being coaxed along by other more vigorous neighbours—and also being compensated for by them. When the spine performs its usual grand-scale activity, each segment contributes a tiny bit more to make up for the stiffer one doing a little bit less.

Most spines have a patchy distribution of stiff links randomly scattered throughout, from the base of the skull to the sacrum. Some areas of the spine are naturally more mobile than others. The neck, for example, is more freewheeling in all its movements, while the low back is much more a fundamental pillar of support. In other parts of the spine, some movements are generous and others meagre. In the thoracic region, sideways bending is never very expansive because the ribs are in the way, but rotation or twist here is very free.

In the low back, all freedoms are kept to a minimum except forward bending. Most of the anatomical details of this part of the spine are designed to help it perform this single most important role. In a sense

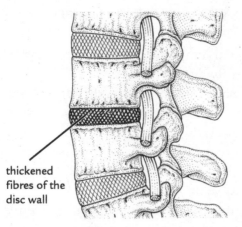

thickened
fibres of the
disc wall

Figure 2.1 The disc of a stiff spinal segment is like a hardened washer between its two vertebrae. This makes the upper vertebra stiffer to palpation pressures.

the main function of the low back is to make the skeleton bendable in the middle and therefore adjustable in height. It is then possible to have your hands, eyes, ears and mouth, indeed the 'thinking' part of your body in positions where it can be the most useful.

It is interesting that the least and most mobile lumbar segments are vertically adjacent. The fifth lumbar vertebra (L5) at the bottom of the stack is the least mobile and L4 immediately above is the most. Although L4 can get stiff, it should come as no surprise that this segment most commonly develops over-mobility problems (see Chapter 6, 'The unstable spinal segment') while L5 is the most frequently diagnosed 'stiff spinal segment'.

A less mobile segment becomes an easy target for trauma because it cannot absorb shock as easily as the rest. Its vertebra cannot roll with the punches and is therefore susceptible to strain. An ill-considered move can leave the problem vertebra locked and twisted on its axis, rather like a screw-top lid of a jar locking down when it is twisted home. If severe enough, the tail of the vertebra (the spinous process) can be visibly out of line with the rest. More commonly though, it reveals a reluctance to be pushed transversely one way compared to the other. The pain from this sort of problem is usually felt on one side of the back only, also manifesting as a 'bruised-bone' feeling whenever the vertebra is touched.

CAUSES OF A STIFF SPINAL SEGMENT

- Unremitting spinal compression reduces disc metabolism
- Gravity squeezes fluid from the discs
- Abdominal (tummy) weakness allows the spine to 'sink'
- Sustained postures accelerate fluid loss and poor varieties of movement prevent fluid replacement
- Chronic protective muscle spasm compresses the problem disc
- Abnormal postures increase neurocentral compression and reduce metabolic activity of the discs
- Injury can rupture the cartilage endplate between vertebra and disc

Unremitting spinal compression reduces disc metabolism

The synthesis of proteoglycans diminishes when a disc fails to undergo adequate pressure changes. As the concentration of proteo-glycans falls the disc loses its ability to attract fluid. The powerful osmotic or 'diffusion' pump gradually loses power, making the disc less efficient at nourishing itself. A starvation cycle sets up as the disc fails to attract water, and stiffens as it dries thus further reducing its mobility and its ability to physically suck fluid in, thus further lowering its metabolic rate. As the stiff disc progressively loses height the health of the entire segment gradually erodes.

Disc de-vitalisation happens more rapidly when the spine is subject to sustained high loads, such as experienced during lengthy periods of sitting, or when the spine is vertically clenched by muscle spasm. It also happens at low loads of the type encountered with lengthy periods of resting in bed. On the other hand, extremes of pressure within physiological range stimulate disc metabolism. Pressure variations within this range are most succinctly delivered by everyday spinal activity.

In addition to the natural squash–suck activity stimulating proteo-glycans manufacture, routine movement also exerts a slight pumping through the discs to move nutritional molecules, particularly the larger ones which have difficulty diffusing. The same convection pumping also helps the disc eliminate waste products and lactates (the by-products of cellular activity) which depress metabolic activity if their concentration rises.

Once a certain level of disc turgidity sets in, the very immobility of the dry, inelastic disc reduces the gradient of the pressure changes *and* the volume of fluid passing through. Both the osmotic and physical pumping engines struggle to circulate fluid and the disc shrinks. Drying and thinning of the discs has immediate ramifications in the cascade of breakdown, making it easy to see why therapy for back problems must focus on disc hydration and nutrition.

Aside from the physiological realities which make disc viability borderline, there are several 'outside' factors that make discs more susceptible to breakdown, thus speeding degeneration of the entire spinal segment.

Gravity squeezes fluid from the discs

All discs lose fluid through the day and replace it during sleep. Gradually, fluids ooze out when we are upright and the spine is compressed, with fresh quantities imbibed again when we are relaxed and horizontal overnight with the segments un-weighted. The slow diurnal pattern of pushing stale fluids out by day and recouping fresh amounts at night is the main way discs circulate fluid, and this stately exchange is only possible because the metabolic rate of discs is so low.

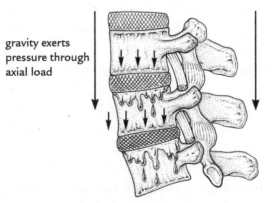

gravity exerts
pressure through
axial load

Figure 2.2 Fluid is pressed out of the spine's discs through the day while we are upright and is imbibed at night as we rest horizontally.

L5 endures the greatest compression as the brick at the bottom of the stack. To tolerate this throughout a lifetime L5's disc starts off thicker than the rest. Over time, constant downward forces can cause incremental fluid loss, for the reasons given above, and low lumbar discs usually flatten faster than elsewhere in the spine. It is common for L5 disc to end up the thinnest.

Intradiscal pressures are highest with sitting because this posture exerts the greatest compression on the spine's base. With long hours of sitting, metabolic activity of the low lumbar discs slows as fluid is relentlessly squeezed out. Discs lose approximately 10 per cent of their fluid within the first two hours of sitting, when the stacked bony segments slowly settle in to closer contact. After this time, fluid loss continues at a slower rate until the osmotic pull exerted by the proteoglycans in the nucleus equals the squashing effect of gravity and fluid loss stops.

The bearing down force of gravity squeezes fluid from every disc in the spine, from the sacrum to the base of the skull. This means we all go to bed appreciably shorter than when we got up. As our muscles relax and our discs swell overnight we are all taller by morning, with our discs primed and plumped up with nutrients, ready to take on another day.

People who sit for long hours may develop progressively more torpid lumbar discs which cannot accept a full quota of fluid overnight. They often feel cast when they rise in the morning, with their back feeling as brittle as a pretzel, and it can take several minutes of being up and about to evacuate sufficient fluid to make movement easier. More significantly, the reduced fluid exchange and gradual disc stiffening is setting up these lumbar discs for breakdown.

Jogging is particularly damaging for low backs because the forces of compression are so much greater. Marathon running eliminates greater amounts of fluid from the discs and overall height loss can be up to 5 cm. Faster running is less dehydrating because the forward stance braces the tummy and draws up the pelvic floor which buoys up the spine and dampens the jarring of the basal discs.

Abdominal (tummy) weakness allows the spine to 'sink'

Strongly contracted tummy muscles play a critical role as a retaining wall when the spine bends. A powerful 'drawing-in' contraction of all three layers of the abdominal muscles creates a higher intra-abdominal pressure which lengthens the spine from within and increases the tensile linkage strength of the disc walls. This provides more security for the segments as the spine goes over, lessening their tendency to shear. Strong abdominal tone also does important things just as you sit there. A co-contraction with the back muscles buoys the spine up and stops the segments squashing down into the pelvis and compressing the lumbar discs.

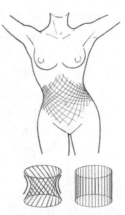

Figure 2.3 Strong drawing in of the abdominal muscle corsetry raises the intra-abdominal pressure and bears the spine aloft which alleviates compression of the lumbar segments.
(I. A. Kapandji, 'The Physiology of the Joints')

When the abdominal retaining wall is weak, it cannot generate sufficient up-thrust to offset the downward forces of gravity. As the girth expands and the belly distends over the belt like a sagging bag of chaff the abdominal contents spill forward, further dragging the spine downwards. As the multi-segmented column ploughs down ever more firmly onto the sacrum the loading pressure on the lower segments increases, adding to the unremitting load that lessens the metabolic activity, particularly of the L5 disc.

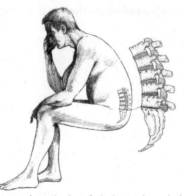

Figure 2.4 During protracted periods of sitting a lax abdominal wall allows the lumbar spinal segments to plough down into the sacrum which compresses the lower discs.

It is worth noting that carrying loads balanced on the head—another time-honoured custom of less developed societies—invokes a superb dynamic response from the tummy and back muscles working in tandem. The abdominals automatically pull in to brace, converting the lower abdomen into a taut, supportive cylinder. The raised pressure within the abdominal cavity creates extra lift for the spine and safeguards the lower segments from excessive compression. We might do well to copy this way of carrying!

Sustained postures accelerate fluid loss and poor varieties of movement prevent fluid replacement

Sustained postures keep the evacuation pump going one way whereas routine activity alternately raises and lowers intradiscal pressures to help suck fluids in and out. In the normal lordotic posture of upright standing the facets bear some load which spares the disc at the front excessive compression and fluid loss. Sustained bending forward disengages the lumbar vertebrae from the compression protection provided by the facet joints and loads the discs more. With sustained leaning over, the facets pull apart and the discs at the front are excessively 'milked'.

Lengthy periods of leaning forward, concentrating the gaze over a small field of vision are commonplace in our work lives, whether it be seated at a keyboard, working on a production line or sowing a paddy field with rice. The forward flexed posture pinches the lumbar spinal segments together at the front and subjects the lumbar discs to sustained compression. High loading is one of the chief factors in paving the way to breakdown of the lumbar spinal segments.

You can see from Figure 2.5 below that the flexed posture of diagram B causes ten times the fluid loss from the nucleus and 50 per cent more from the disc wall than that of the lordotic posture of diagram A. This makes it abundantly clear why sitting in a *relaxed* way with a pillow behind the low back to keep it arched is so necessary, particularly if you have a bad back.

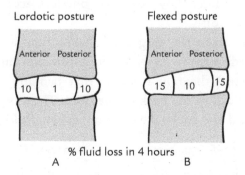

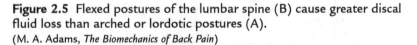

Figure 2.5 Flexed postures of the lumbar spine (B) cause greater discal fluid loss than arched or lordotic postures (A).
(M. A. Adams, *The Biomechanics of Back Pain*)

Repetitive bending is doubly destructive when it co-exists with a meagre variety of 'other' activities to take the skeleton out of its habitual stoop. Sometimes when sitting at a computer we obey our instincts momentarily and stretch back into a wide, non-functional releasing posture but these respites are often fleeting and too infrequent to provide the much-needed pressure alternatives. I believe slumped sitting postures in today's world are the most frequent underlying cause of low-back problems. Sustained low loading also slows the discs' metabolic rate, though in reality, lying horizontal in bed where the discs are not loaded is a much less potent form of disc destruction.

The second important engine for transacting disc nutrition is physical activity. To some extent, discs hydrate themselves by using grand-scale spinal activity to pull the segments apart, thus creating a slight suction effect to entice fluids in from the vertebral bodies. Unfortunately, the relative stability of L5 equates to lack of mobility, which makes it harder for this segment in particular to recoup lost fluids. As the most squashed disc, it is also the most disadvantaged; the trade-off between mobility and stability makes it harder for L5 to keep itself puffed up and buoyant.

This is more significant if the spine is less active for some reason. If you have pain, or more importantly *if you fear movement*, your lower discs flatten faster. With insufficient bending and straightening, the pressure pump is less adequate for shunting fluids in and out and disc nutrition suffers. Low levels of physical activity hasten the decline of lumbar discs.

You will see from the diagram A below that stretching into a lordotic (arched) posture increases fluid diffusion into the front of the discs, whereas diagram B shows that bending pulls a greater quantity of fluid into the back of the discs. Thus, both extremes of bending forward and back take turns at sucking fluid in.

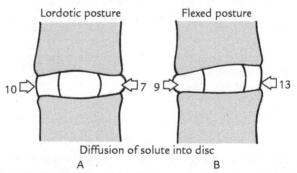

Lordotic posture   Flexed posture

Diffusion of solute into disc

A   B

Figure 2.6 With low-back arching-backwards movements (A) fluid is sucked into the front disc wall. With bending movements (B) fluid is sucked into the back wall and pressed out the front.
(M. A. Adams, *The Biomechanics of Back Pain*)

Movement is the life blood of the discs. Studies in the past have also shown that patients confined to bed for non-back-related reasons lose disc bulk. Height measurements before and after confinement showed that even healthy discs become steadily thinner with protracted inactivity and reduced gravitational stresses. This

is particularly relevant to spinal therapy, where grand-scale movement must deliberately compensate for the reduced proteoglycans
concentration and permeability of the vertebral endplates so typical
of the ageing process.

Lack of physical vigour throughout daylight hours also allows our
muscles and other soft tissues to become less yielding and stretchable,
which makes the spine less free to unravel when it tries to 'grow'
overnight. Thus overall body stiffness indirectly increases the
compression of our discs. When they are not put through their paces
by day they are too stiff to let fluid seep in by night, and for this
reason I like patients to deliberately stretch to decompress their
spine before going to bed.

Healthy discs take in fluid faster than they expel it. It takes
6–8 hours overnight to regain the fluid volume squeezed out the
previous day. Settling discs can gulp in small quantities of fluid by
stretching and twisting during upright hours, though you can make
this up more quickly if you lie on your back and bring your knees to
your chest. This explains why the simple 'knees-rocking' exercise is so
important in your daily treatment plan. 'Squatting' through the day
also helps counteract discal fluid loss, mainly by hinging open the
back of your spine.

Lying draped backwards over a BackBlock passively arches (hyper-
extends) the lumbar spine which potentiates the effect of bouncing
your knees to your chest. The distractive force is the antithesis of
compression and it sucks fluid in as it pulls the segments open at the
front. At the same time, the lower intradiscal pressure stimulates disc
metabolism.

People who deliberately spare their backs by bending their knees
instead, are actually doing themselves harm. It starves the discs of
drink. Certainly, if a weight is very heavy you should lift it like a
weightlifter, but for most daily travails—bending down to get the
detergent from under the sink, reaching up to the top cupboard—you
must make your back do the work, knowing that it is *positively good for
it!* If you persist with a straight back you will be bringing on your own
demise; it makes your back stiffer, the discs cannot feed themselves as
well, you become more fragile and more likely to develop instability at
the problem level.

You see this awkward bending with long-term back sufferers all the time—and sadly they believe they are doing what's best! As they bend over to wipe the table or tune the radio they keep their backs rigid and bend sideways in preference to going forward; otherwise they squat on their haunches with their spines ramrod straight. The oft-cited reason for doing it this way is so as not to 'pop the disc out'. This is also wrong. It has been shown that the intradiscal pressure is just the same with bending (and lifting) whether the knees are bent or not.

Extreme stiffness often makes you misread the cues; you are more likely to interpret bending as a no-go area if your back is painful. But be warned. Protecting your back in this way only makes matters worse. It emphasises the use of the wrong muscles and makes the deep muscles of your spine and tummy so weak they cannot hold the segments together (see Chapters 4 and 6). Most importantly, it makes the problem levels vulnerable to shearing strains as the spine bends and lifts. One mishap when your back is working under duress can transform a benign correctable problem into a tragically incurable one.

Opening the *front* of the discs by lying passively draped backwards with a BackBlock under the sacrum helps drag fluid in. It is a power-fully effective 'anti-sitting measure' which also reverses soft-tissue contracture acquired by hours spent slumped in a chair. Apart from stretching the front wall of the discs it stretches the powerful hip flexor muscles at the front of your pelvis which are apt to shorten when they spend hours in a puckered state. Their tightness creates a lack of give across the front of the groin and down the thighs which makes it difficult to assume a 'normal posture' when you stand up.

In the long term this tightness has adverse consequences because it causes the pelvis to tilt down at the front, which throws the spine out of balance and creates a typical 'bottom out' appearance. Tight hip flexors also make you take much shorter steps because the legs cannot angle back properly at the hips. Again this is bad for the back because in making a decent stride, the spine must twist left and right to compensate for the poor hip mobility.

If the sitting posture is especially slumped it causes adaptive contracture of the anterior longitudinal ligament which runs down

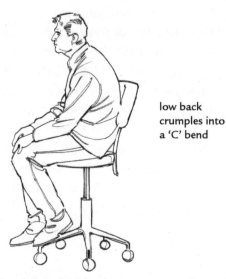

low back
crumples into
a 'C' bend

Figure 2.7 Slumped sitting fails to invoke the tummy muscles and is one of the chief compressors of the spine's base.

the front of the vertebral bodies like a long elastic tape. In healthy circumstances, its role is to limit the backward-arching movement of the spine, but when it adaptively shortens it tethers the spine forward in a hoop, like an over-tight bowstring. People are often aware of their worsening posture, and feel they are being kept stooped, as if their shirt is tucked too tightly into their waistband.

Bear in mind that sitting, or parking one's pelvis on a chair, is a recent and fairly unnatural phenomenon. Many indigenous people still squat rather than use high-backed support. Even though their day may involve running or carrying heavy loads, both of which compress the base of the spine, they can easily disimpact it again come nightfall by squatting to prepare food and eat. Would you see a Masai warrior slumped on a sofa? Frequent squatting exercises form a large part of the self-treatment program.

If you move forward more to the front of the seat, the abdominal wall works more dynamically and gives better support. If you sit slumped your tummy balloons forward, flaccid and inert. If you sit up free of back support you can feel the internal corsetry of your tummy reefing in the retaining wall and making your lower abdomen into a firm, flexible cylinder. It also makes the top of your spine better balanced and freewheeling to work more efficiently over its base.

The best sitting arrangement at a desk is the kneeling chair. Its seat is inclined forward a few degrees which promotes optimal hollowing of the lumbar spine instead of a slumped 'C'. A degree of weight taken through the knees and lower legs on the upholstered cushion minimises that taken through your sitting bones and lower back. Ingenious as they are, it is important to use caution when first using these chairs, particularly if you already have a back problem. The sudden unaccustomed activity of the back muscles trying to hold you up can intensify your pain in the short term. You should start with ten-minute periods per day only and gradually build up the time. It is also important to hasten slowly if you have knee or ankle problems.

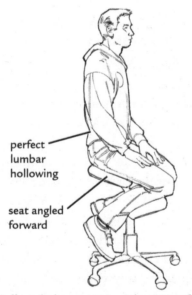

perfect
lumbar
hollowing

seat angled
forward

Figure 2.8 The kneeling chair creates an 'almost perfect' sitting posture where the spinal and abdominal muscles work together to hold you up.

Chronic protective muscle spasm compresses the problem disc

Sometimes, a back over-protecting itself can make a problem snowball. After injury, major or minor, protective guarding by the muscles surrounding the injured segment keeps it out of action until the inflammation subsides. Usually this only takes a day or so and then the muscles relax by degrees, letting in just enough movement to coax the injured fibres to heal. Tentatively at first and then with more gusto, they let the injured segment join in with the rest of the spine. All going well, movement introduced at just the right rate brings the injured segment back to full function with no legacy of pain. But if the injured link never gets going again properly, it will remain an ongoing focus of trouble.

Overzealous muscle spasm can create a stiff link in the spine, even though the original injury was minor. The vertical clench compresses the spine throughout its length, especially at the problem level. Over time, the tissues develop adaptive shortening across the spinal interspace (like the childhood fable where the changing wind fixes a grimace on the face). Thus one vertebra acts like a rusty link in the sleek spinal chain, clonking as the spine goes around and sending out screeches of pain.

There can be a similar outcome if the muscle spasm remains self-fuelling, well after the initial irritability has faded away. This is usually related to subliminal anxieties and a fear of moving the back, when it seems as if the muscles develop a mind of their own. They remain rigidly on guard (the muscles do not pulsate, which is a common misconception), restricting all spinal movement and making everything stiffer and more painful. The cycle is never easy to interrupt and in the self-treatment section you will see how you may have to trick the muscles physiologically into switching off before progress can start.

Pain from simple segmental stiffness has no doubt been around since we evolved to stand upright. Using the hands to get all the vertebrae in line and equally mobile probably went on when we were cave dwellers in a way, believe it or not, which is still appropriate today. On the wall in my clinic I have a print of an ancient Egyptian text with diagrams of human spines being pushed around by the feet, a method I still use today.

However, this book is about self-treatment and I will explain how doing a variety of exercises, using a block of wood, a tennis ball, or perhaps a convoluted rolling pin, can prise individual segments free using your own efforts. All that comes later. Meanwhile, back to what is wrong.

Abnormal postures increase neurocentral compression and reduce metabolic activity of the discs

Deviant spinal alignment can be a potent cause of segmental stiffening. Anomalies can stem from poor postural habits, just as much as from congenital curvatures like spinal scoliosis.

As a rule, spinal segments stiffen more readily in zones where the spine's function alters; where it changes from neck to thorax to low back. These are called the 'transition zones' and they correspond to where 'S' curves (viewed sideways on) change direction. Hence, the common trouble spots in the spine are the lumbo-sacral level (where the spine joins the fixed pelvis), the thoraco-lumbar level (where the thorax becomes the low back), the cervico-thoracic level (where the neck joins the thorax) and the atlanto-occipital joint (where the base of the skull sits on the neck).

A rounded (kyphotic) low back

All transition zones give greater trouble if the natural 'S' bend of the spine deviates too far from normal. However, a fixed kyphotic low back (rounded into a hump instead of a hollow) is particularly troublesome for the neurocentral core because the facet joints at the back are pulled apart by the humping posture and bear no weight at all. This means that the discs themselves take a direct hit of compression, with no mitigation from the facets. The rigid rounding of the low back also obliterates the dynamic bowing forward on the impact of heel strike. Denied the natural sink-and-spring of a working lordosis many spinal levels develop segmental stiffness.

As a further consequence, reduced metabolic activity of the inner 'pinched' sides of the lumbar discs makes them wedge-shaped at the front, thus increasing the severity of the spinal hump. At a time when all the lumbar discs are forced to shoulder a more than usual load they have to make do with reduced nutritional exchange to carry out vital running repairs.

There is another ill effect of lumbar kyphosis: the weight of the upper body is carried too far forward, in front of the line of gravity. Consequently, the spine cannot stack itself with minimum effort and a 'turning moment' (a tendency for the upper spine to slump) comes

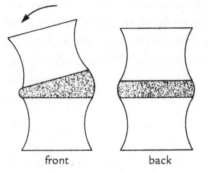

front   back

Figure 2.9 Greater fluid loss and reduced proteoglycans synthesis on the pinched (concave) side of a lumbar discs hasten the development of a more stooped (kyphotic) posture.

about around the thoraco-lumbar junction. With the shoulders stooped, the lumbar segments incrementally slide forward, jamming the spine at mid-lumbar level. A lumbar kyphosis creates a similar tendency for the whole body to tip forward on the pelvis, also contributing to the typical 'bottom out' appearance.

Figure 2.10 A humped lower back (as opposed to lordotic) puts the upper body in front of the line of gravity and causes excessive loading of the lumbar discs.

This is a great source of strain. Pain can emanate from two sites—the bottom and top ends of the lumbar spine—at the same time. Upper lumbar problems often refer pain down lower in the back over the lumbo-sacral junction. When isolating your problem levels it is important not to assume that all your trouble is coming from L5, where the pain is. Both levels must be dealt with if you are to get better.

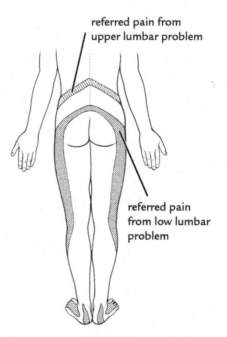

Figure 2.11 Segments of both the high and low lumbar spine refer pain to a similar area across the top of the buttocks.

Running with a humped low back is an exceptional hazard. Even walking can be juddering instead of feline and youthful, with the head no longer tracing an imaginary wavy line along through the air as the spine bounces along with every step.

Even the most youthful spines collapse into a 'C' shape when sitting, though it is better if they keep a proper lordosis, if that is possible with minimal effort. This applies even more when sitting in a vehicle when the added vibration causes greater fluid loss. Accordingly, car seats should have a firm and pronounced upholstered bulge filling out the entire lumbar hollow (to the extent that it feels too much when you first sit down) and the seat should not be tipped down at the back which throws you into slumped sitting on the neurocentral core.

Incidentally, the correct horse-riding attitude is almost the perfect sitting posture for the back to disperse weight. Providing the stirrups are the right length, the low back assumes the optimum alignment to ride out shock and balance the upper body over the pelvis.

A sideways-twisted (scoliotic) spine
Spinal scoliosis is a lateral 'S'-shaped bending-twist through the spine which is obvious when viewed from behind. The lateral angulation of the vertebrae near the apex of the curve pinches them together on one side, reducing metabolic activity and the synthesis of proteoglycans on that side of the discs only. Thus the discs lose viability and their ability to rehydrate, right at the point where they are thinnest. This dramatic one-sided reduction in metabolic activity explains why the discs of scoliotic children deform so quickly and their curves get so dramatically worse during their growth spurt. In effect, the rest of their body keeps growing, except at these focal points on one side of the spine. As growth continues, their body hunches and their chest cage buckles around the spine, as if their growing body is tethered to the Earth by a string of wire from inside the spine.

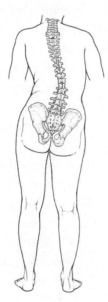

Figure 2.12 A scoliosis is a 'windswept', lateral 'S' bend through the spine when viewed from behind.

Scoliosis can be a congenital abnormality or acquired through the presence of a shorter leg. If a scoliotic curve is relatively stable and does not involve rotation of the vertebrae, the pathology will limit itself to segmental jamming. If however the spine twists as well as deviates laterally, the disorder will include 'facet joint arthropathy' (see Chapter 3). Here, the facets become chronically inflamed in their role of locking the spinal segments together to prevent them twisting off centre. With scoliotic curvature, there is usually a primary curve in the lower back and a secondary curve higher up. The upper curve develops to compensate for the lower one, throwing the spine back across the central line so the head at the top of the column can sit squarely on the shoulders and the eyes can focus to judge distance.

When one leg is shorter, the spine twists one way then the other, compensating in a fairly predictable way. Usually, though not always, the scoliotic lumbar curve will be concave towards the side of the longer leg.

Lateral curves in the spine also develop trouble because the ligamentous shoring of the sides of the column is so weak. Unlike the various structures which keep it stable in the forwards–backwards direction, there is little impediment to the segments sliding sideways off-centre, except the disc wall itself.

The vertebrae themselves develop lateral wedging of the bone, particularly at the apex of the curve. The pinching of the apical vertebra also compresses its intervertebral disc beneath it. In addition, the segments above the apex tend to slide laterally one way and the other way below. As the creeping vertebrae move off-centre their discs also drag sideways, thinning as they go. Thus several discs in the scoliotic curve are thinner, though the apical one more so. As several levels become 'stiff spinal segments' it is easy to see why the distribution of pain from scoliotic spines can be so diffuse.

The jammed apical segment is always the most painful, though several segments may be involved, depending how many times the spine twists back and forth as it goes up. Various spinal segments emitting pain simultaneously account for the wide variety of symptoms typical of scoliosis. There can be neck discomfort (sometimes including headaches), pain in the shoulder blade area (sometimes down the arm), pain at waist level (sometimes referred

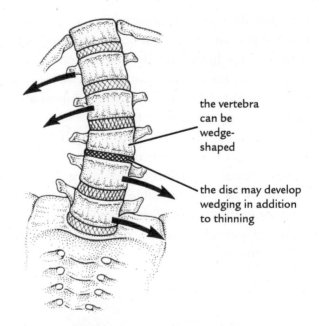

the vertebra can be wedge-shaped

the disc may develop wedging in addition to thinning

Figure 2.13 Poor lateral shoring of the spine allows the vertebrae to creep sideways, one way above the curve's apex and the other below. The apical disc and vertebra are compressed by the spine buckling.

to the groin), and low-back pain (sometimes referred down the leg). With so much pain, mild scoliotic patients in particular are rarely taken seriously and often wrongly dismissed as making a fuss.

Symptoms from exaggerated spinal curves (lordotic and kyphotic low backs) usually emerge during the third decade of life as the internal make-up of the tissues changes and they become more fibrous. Pain from scoliosis, on the other hand, can come on as early as nine or ten years old and be with you for life, getting progressively worse, unless something is done about it.

Injury can rupture the cartilage endplate between vertebra and disc

Unlike the hard cortical bone making up the sides of the vertebral bodies, the vertebral endplates of the roof and floor of the disc are a thin cartilaginous interface. As the spine alternately compresses and off-loads like a concertina during weight-bearing activity, the trapped internal pressure of the discs causes the endplates to belly outwards with the force, like wind billowing up under a tarpaulin. This imparts a lovely buoyant spring to our step as we romp along the pavement, but it also illustrates how weak the endplates are. In fact, they are the weakest component part of the spine.

Vertebral endplates have many tiny holes which allow the diffusion of discal fluids from nearby blood reservoirs in the vertebral bodies. They are pinpoints of weakness and can easily perforate under pressure. It is easily possible to damage the endplates within the limits of normal activity (physiological range) as reverberating shock passes up through the central core rupturing a small vent through the fibro–cartilage interface.

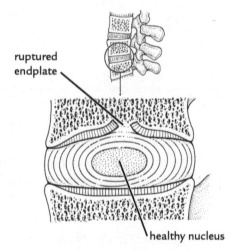

ruptured
endplate

healthy nucleus

Figure 2.14 It is relatively easy to punch a vent through the vertebral endplate during normal daily (physiological range) activities.

The tiny fracture can allow blood from the vertebra to seep into the sterile bloodless outer environs of the disc. In a process similar to a cold abscess, a low grade 'discitis' (or inflammatory reaction) sets up in the disc as it reacts to the invasion of blood. The process may be completely painless because it is contained within the insensitive disc interior but the segment stiffens afterwards. The disc converts to an inert fibrous 'washer' which pads out the interspace and gives the disc bulk, but thereafter it simply acts as a primitive spacer instead of a buoyant tipping ball. It separates the vertebrae well enough but it has none of the highly resilient spinal connector properties which make a healthy disc so brilliant at what it does.

There are many ways you can puncture a vent in the bone, but the most common is stepping off a wall while carrying a heavy weight (particularly on your shoulder) or falling down hard on your bottom (especially if you were expecting a chair to be there). You may also do it wrenching up a sash window which is stuck, or standing up suddenly and hitting your head on an overhead beam (although this is more likely to injure the thoracic part of the spine).

Dramatic as the original trauma may be at the time, it rarely causes severe pain. It jars the spine but discomfort is usually short-lived. Your back may feel stiff and sore for a few days, possibly with leg pain caused by the trauma to the facets. However, you are often aware, in a vague sort of way, that your back is never quite the same again. The injury usually hastens the development of a 'stiff spinal segment', as the disc hardens and narrows.

The process is even less painful when it happens higher up the spine. Often you are unaware of anything wrong until X-rays reveal a typical tear-drop anomaly in the main body of a vertebra called a Schmorl's node. But lower down where the gravitational stresses are higher the sick disc may degenerate much faster, sometimes fast-tracking to entirely different symptoms of 'segmental instability' (see Chapter 6).

THE DISC BREAKS DOWN

As your back gets progressively less mobile the changes to the lower discs become more difficult to reverse. The diminished water content allows the vertebrae to settle closer together and the fibrous walls of the disc bunch down and become less stretchable. When this happens the buoyant nucleus become imprisoned by the hardened mesh walls, rather like the four wing-nuts in each corner of a flower press screwing down the plate. The once-robust bouncing-back nucleus is kept penned inside where it gradually loses life. Instead of being watery and translucent it becomes a yellowy-brown colour

and more viscid (sludgy). It also becomes harder to visually differentiate nucleus from anulus.

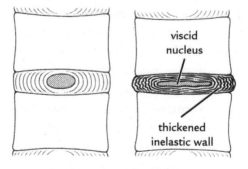

Figure 2.15 A buoyant healthy nucleus imparts a quivering up-thrust to
the spine when it receives compression whereas a degenerated disc receives
load with a thud through the bunched-down disc wall (anulus).

As the disc degenerates, load bearing transfers from the wizened
nucleus to the disc wall. Since the outer wall is tensile rather than
compressive it collapses like a buckling wicker basket as it bears
weight. The bulging disc becomes an inert spacer which manages
to keep its vertebrae apart simply by its bulk. The pulverising forces
of load also create friction between the fibrous layers of the disc wall
(lamellae) creating circular separation and radial splits from the inside
anulus out. Eventually, the wall destruction can lead to 'disc prolapse'
(see Chapter 5) as the degraded nucleus, which has lost the osmotic
cohesion and spreads under compression, starts to burrow through
chinks in the wall to the periphery. However, this turn of events is
uncommon and accounts for fewer than 5 per cent of back problems.

As the disc progressively hardens it ceases to be a stretchable
connector and a band of stiffness develops across the spine's length.
As the spinal segments pull away from one another in movement, like
a beautiful streamlined concertina, the problem link cannot let go.
Fibres of the disc wall ping as stretch is demanded of a link that
cannot provide it and pain signals are elicited from both the chemical
and mechanical receptors. The low back emits a low-grade constant
ache, with sharp jabs of pain whenever you stretch.

It is also difficult for a flatter disc to cope with spinal bending
because there is no buoyant nucleus on which its superincumbent

vertebra can pivot. When it cannot see-saw, the upper segment shears instead and the walls are stretched. Atrophy of the small intrinsic muscles (multifidus) controlling segmental movement also traumatises the brittle link. The weakness comes about partly through reflex inhibition (the less they contract, the less the painful segment is compressed) and also because they have so little to do because the segment is too stiff to move.

Restoring mobility makes a stiffer segment less vulnerable to physical assault, although strengthening weakened intrinsic muscles must be the final step in the rehabilitation process. (Note: strengthening the abdominal muscles is the *first* step.) Remember, whenever spinal strengthening starts it makes your back sore (see Chapter 7) but being aware of this at the outset should dissuade you from taking fright and repairing to your bed. Getting it right with treatment is all about *getting the rate right.*

If a disc continues to stiffen and degenerate with chronic muscle spasm locking it away, it eventually becomes a mass of scar tissue. As it loses its functioning nucleus it cannot rock and pivot to ride out shock waves passing through and it becomes an easily 'knockable' link; a sitting target waiting to get hurt. Instead of a tightly contained ball of fluid, the disc resembles a wedge of compressed carpet which is repeatedly flattened by compression. It develops multiple large-vent ruptures of both endplates, making deep fissures through the disc that are invaded by blood. The pain network can augment with increasing numbers of nerve endings growing into the heart of the sick disc; in effect 'seeking out the pain'. Rare cases such as these are ideal for disc removal and surgical fusion of the segment.

Long before this point, however, the aim of spinal therapy is to rehydrate a drier stiff disc to make the spinal link more mobile. Improved mobility allows the disc to suck fluids through to nourish itself, generating the pressure changes to gear up the metabolic factory within. This healing process is not instantaneous because discs have such a slow metabolic rate but remember, *discs do regenerate.* 'Therapy' is simply about helping discs repair through enhancing nutrition. It is not rocket science; it just takes knowledge . . . and perseverance.

THE WAY THIS BACK BEHAVES

There is a world of difference between the acute and chronic forms of the stiff spinal segment. Strange to say, the more advanced pathologies are often less painful. This is because the segment is so stiff it is almost fused, and where there is little movement there is little pain. At this stage, however, the low internal pressure of the disc means that it retains the inherent risk of knocking loose from its jammed impaction and very quickly becoming unstable. Its lack of intrinsic tensile strength means, quite literally, that if the segment is not stiffly stuck together it is vulnerable to instability.

The sub-clinical phase

In its sub-clinical form this back is hardly a problem. It may be stiff after a long car trip or sleeping in a different bed, but it is never painful. There may be a vague awareness that something is not right with your back, but for years it may come to nothing more than this.

A typical sub-clinical problem is the 'jumpy legs' syndrome, which feels as if two live wires touch when you sit for too long. A lesser form of the affliction is not being able to keep your legs still when sitting, usually combined with a dull sense of pressure in your back. Both conditions are more a nuisance than a problem, and although there is never any pain, it is unnerving and indicates a degree of compression of the spine which is an ill omen for the future.

Sub-clinical disorders wait in the wings to cause trouble. The causing mishap is often trivial, yet brings about an extraordinary pain response. People are often perplexed at how their back became painful so quickly with so little provocation. But they actually had it coming; their back was silently protecting a rusty level and adjusting for its lack of mobility by over-compensating at the levels above and below. For some reason, the ricking incident bypasses all the usual defenses, and unavoidably targets the semi-rigid link.

The acute phase

There is no greater back pain than acute inflammation of a spinal segment. It causes an intense smarting, aching soreness right across the centre of the back which is often too tender to touch. It is often described as screamingly painful with a hot throbbing sensation under the skin, like a boil about to burst. (When this bad it is common to suspect you have cancer.) If the vertebra is locked one way, twisted on its axis, the pain will still be central but focused to one side as well.

At the height of an acute condition your back feels as fragile as a Dresden doll. Jarring it can be so painful it almost makes you sick. Even deviating in your path to avoid a collision on the footpath can make you so weak you almost crumple at the knees. Someone brushing past behind you can make you flinch and automatically move out of the way. Staying upright may be nearly impossible, although pressing the flat of your back into the wall can give relief. Cooking the dinner can be a panicky race against time, until the pain makes you lie down.

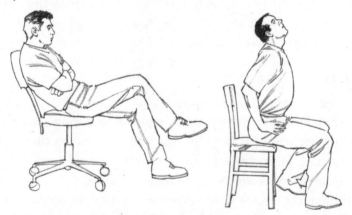

Figure 2.16 With acute segmental stiffness the spine does not like being 'sat on'—you either lie back in a slung-out position, taking weight on the back of your sacrum, or over-arch to relieve the pressure.

Sitting is uncomfortable and your spine does not like being compressed. You constantly shift positions from slumping deeply and resting on the low slung-out back (to take the pressure off the neurocentral core) to perching forward and making it over-arch (so the facets take more weight). Either way, your back is uncomfortable again within moments. Even lying down can be painful because your back feels too tender to take the pressure. The locked vertebra seem pushed up by the muscles, as if you are lying on a stone.

> **Acute palpation**
>
> When I use my hands to palpate, the vertebra can be highly sensitive to light pressure but surprisingly less so to deep. En masse, the vertebrae feel bunched together, like a row of beads threaded too tightly on elastic. Sometimes the segments feel hard to depress like piano keys where the spring underneath is too stiff. Often the surrounding tissues have a puffy, water-logged feel with a deep inflammatory heat smoldering up from below.

What causes the acute pain?

With acute segmental stiffness there are multifarious reasons for the pain and there is plenty of it.

The injury which first hurts your back can be likened to a ligament strain of the disc. It is similar to spraining an ankle though on a smaller scale. Chemical toxins are released when the fibres are damaged which irritate 'nociceptors' or free nerve endings in the disc wall. Messages are relayed to the brain from the injured part, which are interpreted as pain.

Having said this, the disc is poorly sensitive to pain. Only the outer layers of its wall have a nerve supply which makes it unlikely to bring about intense pain on its own. This rather points to the cramp of local muscles in spasm and the accumulation of waste products around the injury as additional sources of pain.

Although the muscle spasm is protective it can be too unrelenting. Unabated, it physically jams the segments together and increases the pain coming from the sore interbody joint. Mechanical receptors with bulbous or globular nerve endings in the disc wall are stimulated by the excessive compression. These are situated between the fibres and are sensitive to physical distortion. They are flattened when the disc is flattened and this too is perceived as pain. Sometimes this stimulates more protective reaction from the muscles and the painful cycle intensifies.

Probably most of the pain comes from the intense vascular engorgement around the injured part when the muscles prevent free movement. The disc stays compressed and the circulation of blood becomes sluggish because there is no pumping action from free movement to sluice it on. Pain comes from the physical engorgement of the neighbouring pain-sensitive structures and also from the rising concentration of waste products in the stale blood.

This engorgement is a potent source of discomfort. It accounts for the steadily increasing pain, several hours after injury, just like when you twist an ankle. There is a wrench of pain when the mishap first happens which then passes off, but several hours later the pain worsens. As the joint gets stiffer and more tense with swelling, there is a frightening inexorability about the way the pain gets worse.

The sub-acute phase

In this phase, the discomfort from the low back is more bearable. The lumbar spine feels permanently clenched, with peaks of stinging pain or twinges whenever it gets tired. It suddenly gets uncomfortable being in one position for too long and is only relieved by moving about. Rubbing with the flat of the hand is a relief, although direct pressure on the vertebra feels tender, as if the bone itself is bruised. The aching stiffness can be relieved by heat, sometimes so hot it works like a counter-irritant. (Using a hot-water bottle is common but often leaves a mottled discolouration of the skin which takes years to fade.)

Movement is painful because the muscles are tight and the back fails to let go as you try to push through their clench. Standing becomes more painful as the spine becomes more cast. It worsens into brittle impaction if you stay there and then it hurts to sit down. The spine cannot pull its segments apart to get itself rounded, and folding up to get into the car after standing at a cocktail party, for example, can be excruciating.

From the sub-acute phase, the problem can pitch back into acute flare-ups when the jammed link is disturbed or it can become more subdued and move into the chronic phase. It typically see-saws between the two with painful spates and remissions, and shorter respites in between. At this stage, avoiding hurting your back can become life's obsession. You take the long way round doing every-thing, just to avoid setting it off, and often the whole family must come to terms with accommodating your back.

Sub-acute palpation

When I palpate with my hands the segments have usually lost their bunched-together feel, but there will be a tubular rigidity across a local section of the neurocentral core. The spine feels brittle, as if the cotton reels are welded together and the surrounding tissues may have the rubbery feel of longstanding inflammation. With the flat of my hand I feel the drag of moisture on the skin which indicates a deep seated, low-grade inflammation. After manually mobilising the vertebrae the skin often flushes up red with the blood rushing to the surface. (The degree of redness gives some indication of the degree of inflammation.)

The chronic phase

The chronic phase of segmental stiffness makes you feel years older than you are. Rather than frank pain you have a deep, aching, armour-plated stiffness across your low back. Arching backwards gives relief, but bending forward is always awkward and stiff; your back feels so rigid you sense you shouldn't do it. It is difficult drying your toes and putting your socks on in the morning but activity gets easier as the day goes on. You feel creaky getting out of a chair or car and you often have to winch yourself straight before moving off.

Chronic palpation

Delving around in a spine like this, it seems the fire has gone out leaving only the cold embers. There is no soupy inflammatory feel of the tissues because they are so inert and lacking in juice they barely react. The vertebral column feels like a semi-rigid mass with thickened bars of bone across the segmental junctions. Often the bone of the vertebrae feels enlarged, like barnacles encrusted on an anchor chain, and the segments have a rock-hard blocking resistance to passive, gliding pressures from me.

What causes the chronic pain?

When a stiff spinal segment passes from its acute to chronic phase the pain comes about for different reasons.

During the course of everyday activity most of the weight through the segment is taken by the disc wall. This causes the wall to pucker and bulge, and the bulbous mechanical receptors hidden between the fibres are stimulated. If fibres are pulverised and broken by excessive compression the chemical receptors will be activated by the toxins of injury. Thus the brain receives two different pain messages.

Pain can also be provoked by the stiff disc wall being stretched. When the disc is inelastic and not free to pull apart with the other segments, the mechano-receptors pick up the lack of give in the fibres and interpret it as pain. If fibres are broken through being stretched beyond their limit, the chemo-receptors pick up the toxins released by the injured tissue.

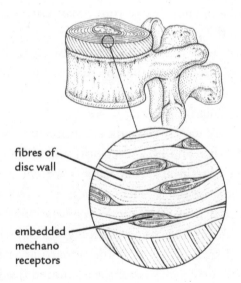

fibres of
disc wall

embedded
mechano
receptors

Figure 2.17 Bulbous mechano-receptors between the fibres of the anulus pick up both compression and tensile stretch of the wall.

Pain from tissue tightness can also be registered in the other ligaments which help the disc hold the vertebral space together, in particular the posterior longitudinal ligament. As the problem disc drops in height during the degenerative process there is adaptive shortening of the ligament across the interspace.

The posterior longitudinal ligament in particular has a highly sophisticated nerve supply and, once a tight band develops, the ligament will register pain—just like the disc only more so—when provoked by stretch. Because it runs right down the back of the vertebral bodies, its tightness makes bending forward particularly painful.

WHAT YOU CAN DO ABOUT IT

Aims of self-treatment for segmental stiffness

The overriding mission with segmental stiffness is to reduce the compression and add buoyancy to your basal spinal discs. This frees them to undergo the pressure variations so essential for stimulating proteoglycans synthesis and sucking water through. When they can circulate more fluid to bolster nutrition the disc metabolism picks up and, with that, disc maintenance and repair.

Mobilising and decompressing a stiff spinal segment is much more straightforward with the chronic condition because your back is barely sore. It is achey and bone-deep stiff and although exercising may stir it up, the benefits are immediately apparent.

Direct loosening of the segment is best achieved by a physical therapy professional who manually loosens the link, although the pressure-change therapy or 'mechanobiology' is only achievable by you, using the BackBlock and doing squatting exercises. Both techniques immediately make use of the newfound freedom, while at the same time introducing wide variations in pressure to torpid lumbar discs.

When the back is acutely inflamed, muscle spasm is the wild card which complicates everything. Although its role is ultimately protective, much of the overall pain picture with problem backs can be attributed to it locking up the spine and holding it too rigidly. Not only does this compress the segments more, it causes more pain from the muscle clench itself. Both factors make your back screamingly painful and also extremely unpredictable from one moment to the next.

It is very important that your self-administered mechanobiology does not increase the resting levels of muscle spasm, which it is apt to do. If, in practice, your decompression exercises are making you sore then you have to lay off for a while (usually a week to ten days) until the back settles. You never use the BackBlock while the back is in the acute phase but you must resume at some stage, if you are to go forward.

Getting rid of the muscle spasm is always a first priority, although all the techniques to do this help loosen the segments as well. Spasm in the muscle is eased initially through stretching, by lying on your back and very gently bouncing, or oscillating, your knees to your chest. You can break up the brittle castness of your spinal segments by rolling back and forth over the low back, and very importantly, you 'switch off' the back muscles by strengthening their opposite number (the tummy muscles) through making the lower abdominals exercise strenuously with reverse curl ups.

A typical self-treatment for acute segmental stiffness

Purpose:
Ease muscle spasm to relieve compression on disc, disperse inflammation, and strengthen tummy muscles to switch off over-activity of the erector spinae muscles and relieve compression on disc.

Rocking knees to the chest
(60 seconds)
Rest (with knees bent for
30 seconds)

Reverse curl ups
(five excursions)

REPEAT BOTH EXERCISES 3 TIMES

Remain resting in bed, only getting up to use the lavatory. Use medication of NSAIDs (anti-inflammatories), painkillers and muscle relaxants as directed by your doctor. Repeat regimen of exercises in bed every 2 hours. See Chapter 7 for descriptions of all exercises and the correct way to do them.

For how long? This phase usually lasts anything from two or three days to a week. You can advance to the next treatment stage when it is easier to sit up and turn over in bed. Remember, fear holds you back. Relax in bed and stay floppy. If you jar your back, relieve the pain by rocking your knees to your chest and resting with your lower legs supported on pillows. Try to relax and not think about your back or the pain.

A typical self-treatment for sub-acute segmental stiffness

Purpose:
Ease muscle spasm to relieve compression on disc, disperse inflammation, strengthen tummy muscles to switch off over-activity of erector spinae and relieve compression on disc, break up spinal impaction caused by muscle spasm, and stretch facet capsules to relieve spinal impaction at the base.

Rocking knees to chest
(60 seconds)

Rolling along spine
(15–30 seconds)

Reverse curl ups
(five excursions)

Squatting
(down twice for 30 seconds)

REPEAT ALL FOUR EXERCISES 3 TIMES

Medication of painkillers and NSAIDs only. Two 20-minute rest periods each day. Repeat the regimen morning and evening every day on a folded towel on carpeted floor.

For how long? Usually, you may progress through this period in ten days to two weeks. It cannot be hurried and you need regular rest periods between exercises. Progress to the next regimen when the pain is intermittent and bending less painful.

A typical self-treatment for chronic segmental stiffness

Purpose:
Ease muscle spasm to relieve compression on disc, disperse inflammation, strengthen tummy muscles to switch off over-activity of erector spinae and relieve compression on disc, break up spinal impaction caused by muscle spasm, stretch facet capsules to relieve spinal impaction at the base, decompress the spine to promote disc regeneration and repair, and re-establish intrinsic muscles' segmental control.

Rocking knees to chest
(60 seconds)

Rolling along spine
(15–30 seconds)

Reverse curl ups
(five excursions)

Squatting
(down for 30 seconds twice)

BackBlock routine   Step 1   (60 secs)

Step 2   (30 secs)   Step 3   (15 times)

REPEAT ALL FIVE EXERCISES 3 TIMES

Toe touches
(down and unfurling up
to vertical twice)

Repeat program every evening before going to bed. Continue NSAIDs if back is too sore to exercise.

For how long? This phase may continue indefinitely but is prone to relapse if you do too much. Always sprinkle your day with toe touches, perhaps two or three morning and afternoon and always squat after lengthy periods of sitting. If back remains sore, stop BackBlock (only) and revert to sub-acute regimen for 2–3 days, concentrating on the spinal rolling.

3

Facet joint arthropathy

Facet joint arthropathy is wear and tear of the facet joints. After segmental stiffness, I believe it is the most common cause of low-back pain.

WHAT IS FACET JOINT ARTHROPATHY?

The term 'arthropathy' covers the wide range of this disorder, from fleeting joint sprain of the capsular ligaments, right through to frank arthritis of the joint.

Breakdown of the facets comes about in several ways: when the disc between two vertebrae flattens, causing the upper vertebra to settle down on the one below and jam the back compartment, just as letting air out of a car tyre makes it run along on its rim. It also develops when the facets have to overdo their preventative role, either restraining forward bending, or preventing twist of the lumbar segments.

The lumbo-sacral facets are particularly taxed when the sacrum is permanently tipped forward, to cause a deep hollow in the low back (a pronounced lumbar lordosis). Then the opposing surfaces, of the lumbo-sacral facets particularly, are forced to butt up against one another to stop the spine slipping forward off the sacrum. A leg length discrepancy can also affect several facets in the low back (and even into the thoracic spine and neck if the difference is great enough) as the lateral dip in the sacrum encourages the lumbar segments to slide sideways and twist.

However, attrition of the joints caused by these background factors may take years to bring facet pain on. Trouble usually comes to the fore when lurking inflammation is provoked by additional small-scale injury. The slightest ricking or twisting incident, such as slipping on a shiny floor when you are turning to talk, or turning over in bed, can put a match to smoldering trouble and set it blazingly alight.

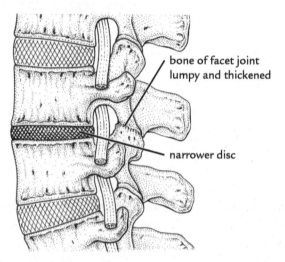

bone of facet joint
lumpy and thickened

narrower disc

Figure 3.1 Facet arthropathy can begin as a fleeting capsular strain—similar to any joint sprain—and eventually becomes frank arthritis, with bony outgrowths (osteophytes) developing around the margins of the joint.

I believe capsular inflammation of the facets is very common and a potent source of back pain. Because the facet capsules play a substantial role in protecting the facets they are easily sprained. They act as tough elastic sleeves to absorb or dampen all jolts passing through the facets, thus sparing them bone-to-bone bruising.

The facets are the most likely part of the spine to pull apart and come undone; in effect their sliding-apart freedom at the back of the spine is also their Achilles heel. The lumbar spinal segments have great difficulty controlling the first imperceptible degrees of bend-and-twist because there are so few diagonally criss-crossing ligaments and muscles to control the action (see Chapter 4). We depend on the disc being primed to give us this stability but the facet capsule too has to do what it can to be all things to all movements. The controlling of the joints and sheltering them from impact mean the facet capsules are constantly in the line of fire.

In most respects the capsules are admirably equipped to cope. They are incredibly elastic and strong with a rich blood supply to keep up running repairs. However, their onerous protective role can cause inflammation and swelling which may irritate a nearby spinal nerve, creating severe leg pain.

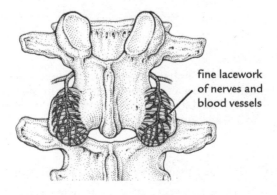

fine lacework of nerves and blood vessels

Figure 3.2 Facet capsules are extremely strong and richly served with both blood and nerve supply.

The facet capsules have a prolific nerve supply to pick up any affront to the joint and relay it to the brain. However, the very sophistication of the nervous network may be part of the making of a back problem; the super-sensitivity of the facet capsule may make it slightly too ready to invoke a reaction from its muscles.

This enthusiasm to protect the joint after it has suffered a strain can make a major problem out of a minor one. Thus a tiny wrench to your back when wrestling with a suitcase can lock you up for weeks, when it should be gone in a day. On the other hand, supreme sensitivity from long-term joint pain may cause the muscles to reflexly under-act, thus causing another set of problems.

When the joint guarding is too attentive it jumps into action at the first sign of trouble. The nervous mechanism lights up and the muscles (local fibres of multifidus and the long erector spinae muscles if the reaction is extreme) clench to protect it more. The prolonged muscle contraction can cause more inflammation, by slowing the

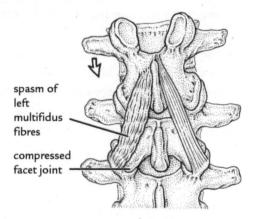

spasm of
left
multifidus
fibres

compressed
facet joint

Figure 3.3 In the acute phase, multifidus develops a holding contraction (spasm) which can keep a facet joint engorged. In the chronic phase, multifidus reflexly inhibits, the 'under contraction' rendering the facet vulnerable to extraneous movement.

flow of blood through the capsule. As the joint becomes more engorged it sends off more messages of pain to the brain and the protective cycle intensifies.

This over-reaction explains why it is so important to work through any minor discomfort and not allow it to get the better of you. It may also explain why the muscles stop acting altogether in cases of chronic facet pain. We know that multifidus in particular is under-active when the back has been painful for a long time and this may be another type of defense mechanism—this time deliberate under-activity to spare the sensitive joint added compression. Although the back may be more comfortable in the short term, in the long term there is a risk of introducing instability to the segment.

Whereas mild capsular strain is relatively easy to acquire and fairly universal, true 'arthritic' change of the facets' bony surfaces is less common and may take years to evolve. Furthermore, I believe that even when arthritis is painful, its origins (certainly in the early stages) are 'capsular' rather than 'bone', at least until it gets to the point of actual bony destruction.

This may explain why treating advanced arthritic facets with the hands—in the way physiotherapists, osteopaths and chiropractors and some masseurs do—often relieves the pain. The accessing and handling of the tissues around a problem joint often interrupts the inflammatory cycle in a way which cannot be explained as doing things to the bone. Unlike the bone, capsular changes are more reversible and respond quite quickly to the comfort of treatment pressures.

Diagnosis by manual palpation

Just as with early stiffening of a spinal segment (Chapter 2), early capsular changes of the facets are only detectable via palpation with the hands, particularly the thumbs. Although the pain from an acutely inflamed joint may be crippling, routine imaging may be equivocal. In the same way, a twisted ankle is unlikely to show anything on a picture; the best way of telling if a facet is troublesome is by sensing whether it will move, by the way it feels, and by the way it responds to being handled.

Although not necessarily relevant in a self-help book such as this, it is interesting that human hands can tell such a lot about a facet problem. With early—otherwise undiagnosable—trouble with a facet the capsule feels like a dome of pulpy swelling under the skin when the thumbs probe in about 1.5 cm out to the side of the spine. If the problem is longstanding the capsule has a thickened leathery feel, caused by the chronic fibrosis, whereas a normal one will feel like nothing at all; health is conspicuous by its inconspicuousness. Experienced hands can also feel if there is over-activity of multifidus (it feels twangy, like an over-taut mini trampoline) or under-activity (when there is a leathery hollow where multifidus should be). Careful reading of MRIs can also reveal local atrophy of multifidus and a higher fat content, indicative of under-activity.

Once there are bony changes, palpation through the hands can tell how mobile (or immobile) the joint is and whether there is any crepitation, or grinding in the joint, as the joint surfaces move past one another. Again, this is not a finding which can be picked up any other way than manually, although in the neck you can often hear very loud chafing from the joints as you turn your head. These palpatory findings are the preserve of professionals but they give a deeper understanding.

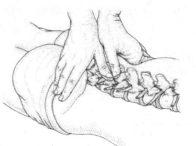

Figure 3.4 There is no better tool than the human thumb to feel facet joints and gauge their function.

Once the joint surfaces become involved in the degenerative process, the changes stand out more clearly, even on X-rays. Because bone is radio-opaque, bony erosion can be seen, as can excess bone growth around the margins of the joint.

The pain is different in the various phases of the condition. Capsular pain is angry and mercurial, sometimes causing pain in different parts of the leg as the nerve is compressed by the swollen capsule, with changes in the body's position. By the time it has progressed to bony breakdown it is a deep-seated, gnawing ache which goes through like a meat cleaver in the side of your back, or across to the other side too if both joints of the segment are affected.

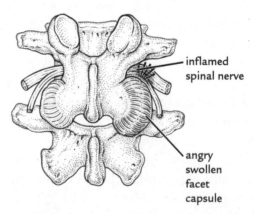

Figure 3.5 Capsular inflammation and swelling can readily irritate a nearby spinal nerve root.

There has *never* been a clear correlation between imaging and degrees of pain. Sometimes the films show nothing at all and yet a patient is genuinely crippled by pain. Somebody else's images can look like a snow storm, with white, roughened bone surfaces and strange knobbled outgrowths around margins of the joints, yet there has never been a day of pain. Hence the maxim of our practice: treat the patient not the pictures.

CAUSES OF FACET ARTHROPATHY

- Disc stiffening allows the facet capsules to tighten
- Disc narrowing causes the facet joint surfaces to override
- A sway back causes the lower facets to jam
- Weak tummy muscles can jam the facets
- A shorter leg invokes a greater restraint role of the facets

Disc stiffening allows the facet capsules to tighten

The earliest form of stiffness of a facet joint may be nothing more than a fleeting protective spasm of its overlying multifidus muscle. A fluke odd movement or an awkward posture for a while, and the muscle locks the joint temporarily to protect it. It is noticeable as a sore tight patch beside your spine which feels like a crimped link but which passes off within a day or two.

Stiffness of the neurocentral core at the front of the segment can translate across to the strong capsular ligaments at the back. Because of the strength and toughness of their fibres, these ligaments are the first to lose stretch as the mobility of the front compartment declines. Even before obvious loss of disc height, disc immobility can greatly reduce the freedom of the facets.

In their less yielding state the capsular structures are much more vulnerable to injury. Being repeatedly yanked by everyday movement amounts to micro-trauma so that more and more capsular fibres are torn. On a microscopic scale there is oozing of blood and lymph into the interstitial spaces (between the fibres) where it lies about and gradually solidifies. This is scar tissue or adhesions. As the capsule

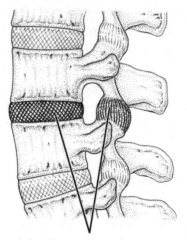

Figure 3.6 Stiffness of the disc can translate across to the back compartment at the same level, making the facet capsules tight.

becomes increasingly cobbled by adhesions it stiffens and the facet joint underneath loses play.

This augments the chain reaction fuelled by the facet joint as well. From now on, both the back and front compartments of the segment contribute to the poor freedom of the upper vertebra articulating on the lower one. As the intervertebral disc loses vitality the stiffness of the link can be felt from the outside. It is less yielding to pressure as you roll back and forth over it and often feels like a plug of cement in a rubber hose. (Sometimes you can feel a stiffer facet to the side of the spine by tipping from side to side.)

Disc narrowing causes the facet joint surfaces to override

A disc of normal height provides a natural 'clearance' for the working facet joints of the back compartment which dictates that they bear only approximately 16 per cent of load. As a disc loses fluid—and height—the upper facet surface can ride so far down the lower one that it digs into the neck of bone at the base, making it take up to 70 per cent of the load through the segment.

Facet joints are not designed to bear load on this scale and breakdown picks up apace. The lining membrane of the capsules secretes greater amounts of synovial fluid to keep up the embattled

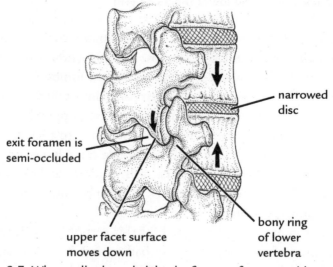

exit foramen is
semi-occluded

narrowed
disc

upper facet surface
moves down

bony ring
of lower
vertebra

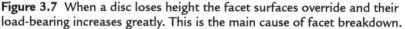

Figure 3.7 When a disc loses height the facet surfaces override and their load-bearing increases greatly. This is the main cause of facet breakdown.

lubrication and sluice the joint interfaces clean. The synovial fluid also carries large phagocytic cells which surround and devour each tiny particle of cartilage, and in this way the joints are kept running as best they can. The greater the internal friction of the joints the more synovial fluid is pumped in, like tears flooding the eyes to get rid of grit, until eventually the excess fluid becomes a problem in itself.

The trapped fluid can cause pain, but the tension of the joint may also invoke reflex contraction (spasm) of the multifidus muscle which lies immediately over the top. As the muscle fibres shorten, the joint is held more firmly and compressed, increasing the pressure from the trapped fluid inside. Although this protective response has not been documented, I suspect it may account for the rapid alteration in the feel of a tense facet when it is touched by probing thumbs. The typical dome of capsular swelling can subside so quickly it feels as if a release valve has let the fluid escape. This may be multifidus letting go, allowing the joint to move freely, thus evacuating its fluid. I am always pleased by how quickly mobilising can bring this about and alleviate pain.

A sway back causes the lower facets to jam

If the angle of the sacrum tips forward more than its average 50 degrees the spine is forced to hollow more as it arches back to the upright again. This causes inordinate wear of the lumbo-sacral facets. In some people the sacral angle can approach almost 90 degrees (with the sacral surface nearly vertical), and the two opposing surfaces of the L5–S1 facets remain permanently jammed to keep the spine hooked on to the sacrum. In effect the whole spine hangs on to the pelvis at these two bony hooks, like sash window catches, and this takes its toll. The temporary sway back of advanced pregnancy causes pain for a similar reason.

The facet joints are not designed for this sort of heavy-duty wear. The closedness of the joint surfaces is bad enough but their excessive grinding is much more destructive. With a normal lordosis the facets are in similar contact only when the spine is fully bent forward, although the posterior ligamentous lock, which comes into force when the back is fully rounded, shares some of the load. When an overly lordotic spine bends forward, the ligamentous lock cannot operate as effectively because the lumbar hollowing puts it on the slack (stressing again why it is so important to bend and lift with the back humped, not arched).

Excessive use of the facet stop-ramp puts the facets under all-day duress and abrades a continuous spume of cartilage off the joint surfaces. This gritty debris floats around in the joint space, acting as a micro-abrasive which scours down the residual cartilage surfaces even faster.

In extreme cases of lordosis, the upper facet surfaces override so far down the lower ones that the tips of the upper bony pillars come to rest at the base of the lower ones. The two fine prongs of bone projecting down from the vertebra above (the front of which provides the articulating surface) are no match for a disc when it comes to spreading load, and breakdown escalates.

Along with the bony changes, there can be marked soft tissue contracture of the joint capsules, in effect creating a bow-string which keeps the spine over-arched. As the fibres shorten and the roominess decreases, the joint surfaces become so close-packed they find it hard

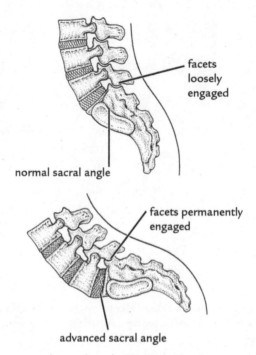

facets
loosely
engaged

normal sacral angle

facets permanently
engaged

advanced sacral angle

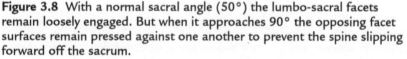

Figure 3.8 With a normal sacral angle (50°) the lumbo-sacral facets remain loosely engaged. But when it approaches 90° the opposing facet surfaces remain pressed against one another to prevent the spine slipping forward off the sacrum.

to pull away from each other to let the spine hump forward. This greatly reduces the bending freedom of the spine.

The backward arching action of the spine can become even more limited as the upper bony tips dig into the base, even jacking the inter-body joint open as the spine tries to arch. If there is simultaneous impact of the foot hitting the ground as the back arches, the bony ring below can break. We see this as stress fractures of the spine with fast bowlers in cricket.

With the spine resting long term on the facets rather than the disc, the vertebral body can demineralise by being stress-shielded from normal gravitational forces. This is thought to be one of the processes whereby the vertebral bodies become osteoporotic and undergo spontaneous crush fractures. The same process can also take place in the absence of extreme lordotic postures, simply through the disc losing height and shunting more load onto the facets.

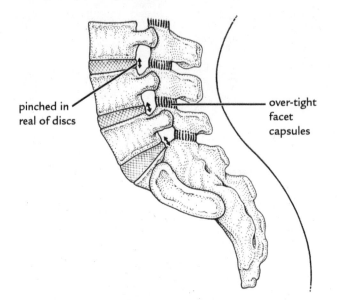

pinched in
real of discs

over-tight
facet
capsules

Figure 3.9 With extreme lordosis the permanent overriding of the lumbar facets results in adaptive shortening (soft tissue contracture) of the bulky facet capsules.

When the sacral angle remains marked over a period, there is adaptive remoulding of the bone of the lower facet surface to create a bony impediment to the spine slipping forward. This is similar to the way an unstable joint sprouts more bone around its edges to keep

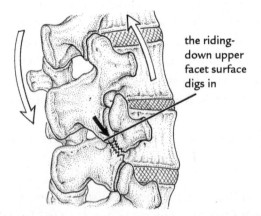

the riding-
down upper
facet surface
digs in

Figure 3.10 Heavy impact with the ground while the back is arched can fracture the bony ring at the base of the facet joint. This is a common cricketing injury.

the facet in joint (see Chapter 6). Nature comes up with an ingenious way of making these joints more secure. A bar of bone forms across the lower facets to bolster their stop-ramp function, like bolting a steel bar across railway tracks. This minimises forward trespass of the spine on the sacrum.

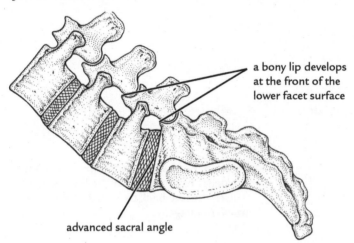

a bony lip develops at the front of the lower facet surface

advanced sacral angle

Figure 3.11 With extreme lumbar lordosis the bone remoulds to throw out a ledge at the front of the facets to stop the spine sliding forward off the sacrum.

Strangely enough, the occupations which cause most facet trouble put the back into the fully stooped posture rather than an over-arched one. Shearers and farriers often spend hours bent double, with their facets fully engaged at the top of the stop-ramp. Initially this causes a ligamentous strain by stretching the fibrous capsules of the facets. Later on it causes bony degeneration from the grinding of the facet surfaces perpetually locked up against one another to keep the spine hooked onto the sacrum.

Weak tummy muscles can jam the facets

Weakness of the tummy can bring about a similar lordotic effect—but at least it is more under your control. As the tummy muscles weaken they often passively lengthen at the same time. As they stretch, they allow the front of the pelvis to tip down, causing a pronounced

Figure 3.12 A weak abdominal wall allows the pelvis to dip down at the front which causes the low back to arch and the facets to over-engage.

hollow in the low back. This causes the lumbo-sacral facets to engage, spending most of their time working as stop-ramps to prevent the rest of the spine sliding forward down the sacrum.

This is part of the explanation for backache which goes with ordinary old fat-tummy obesity. As the tummy gets bigger and more weight is carried in front of the line of gravity, the lower back goes into an even deeper hollow as you over-arch backwards to balance the weight out front. As the sacrum tips down, the lower lumbar facets lock into apposition to keep the spine on the pelvis. Simple abdominal strengthening is very effective at decreasing lordosis and is an important part of treating this problem.

A shorter leg invokes a greater restraint role of the facets

Most of us have one leg shorter by a millimetre or two, which I believe is a common cause of back pain. A shorter leg places great strain on the low facets to hold the spine locked in place on the pelvis. The greater the discrepancy, the greater the diagonal sideways-and-forwards trespass of the spine down the sacrum.

The spine slips forward as well as sideways because the hip joint lies in front of the centre of gravity. Thus on the side of the shorter leg the pelvis dips down at the front as well as laterally on that side. The combination of the two aberrant tendencies causes great wear and tear on all lumbar facets but the lumbo-sacral in particular.

The mechanics of distortion are more complicated than first imagined because the vertebrae then rotate around a central axis of movement, as well as slipping diagonally across the sacrum. A new centre of movement comes into effect as the facet on the downhill side engages and the vertebra swings around this new pivot to twist further. All this makes for complicated movement of the low lumbar vertebrae when one leg is shorter.

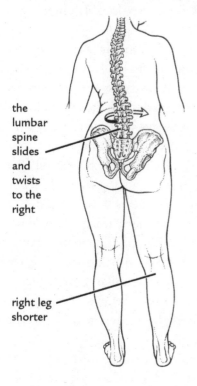

the lumbar spine slides and twists to the right

right leg shorter

Figure 3.13 With a shorter right leg, the spine tends to shear forward incrementally and sideways to the right. As the right lumbo-sacral facet permanently engages, the spine swings around it as it twists to the right.

Furthermore, the hip joint of the longer leg develops trouble too. It acquires a tightness at the front because that leg always stands with the knee bent. This drops down that side of the pelvis and equalises the sit of the sacrum. Permanent contracture of the front of the hip of the longer leg makes length of stride uneven. Because it cannot angle back fully, the spine has to compensate while walking by twisting more to that side to even up the steps. With all of us taking thousands of steps per day, you can see the spine must repeatedly twist one way as we walk. Ultimately, there is a cocktail of irritants where it joins the sacrum.

It is not really possible to be pedantic about which facet will be the most painful at the lumbo-sacral level, although there is a rule of thumb that the facet on the side of the shorter leg will be painful in the early years while the side of the longer leg will be more painful later on.

With extreme discrepancies, the downhill facet ingeniously remoulds, similar to the way it does in an over-lordotic spine. In what are termed 'wrap-around bumpers', the lower facet surface develops bony outgrowths to stabilise the lateral trespass of the upper vertebra. Although these excess knobbles of bone can look frightening on X-rays, they are not indicators of pain.

This 'facet tropism' or asymmetry of the lumbar facets is an issue which causes great excitement in academic circles. However, I believe the high incidence of differing leg lengths explains the prevalence of tropism. It is the readiness of the spine to adapt to the anomalous sit of the sacrum (especially if there is excessive lordosis as well) which accounts for the dissimilarity between two paired joints.

Tropism has been unearthed by researchers who have been quick to point out its strong link to back pain. As a shop floor clinician, I see things from the other perspective: the facets *become* different to help reduce back pain when the legs are born unequal in length, rather than the facets themselves *being* different at birth and causing back pain. This might banish your despair on being told you have back pain because your lumbar facets are dissimilar. I see leg-length inequality as a highly significant factor in the genesis of back pain— although conventional wisdom decrees that differences of less than 2 cm are unimportant—a verdict I deplore.

Correcting leg length discrepancies by using a heel raise in the shoe of the shorter leg is an early and mandatory part of self-treatment. Only an approximate adjustment is necessary; it is better not to jack up the difference to the nearest millimetre. Since you have coped for so long with one leg shorter, the discrepancy should be minimised rather than fully corrected—or it simply adds another set of strains to the pre-existing ones.

Golf clinic

As a result of the rotation involved, golf can often traumatise the facet joints because of their role in limiting spinal twist. As a right-handed golfer swings to the left, the row of facets down the right side of the lumbar spine butts up against each other like doors against a door jamb. The facets down the left side of the spine pull apart. The forcible closing on one side and wrenching open on the other can cause breakdown.

Serious golfers will cause less damage, and get a better swing, if they take the twist higher in their back into their thorax. From waist level up, the facet alignment is different and no longer stops the vertebrae swivelling. If golfers employ the 'flying elbows' technique of taking the twist in an arching spiral up the back, they will not only avoid damaging the facets but hit a better shot.

THE WAY THIS BACK BEHAVES

The acute phase

The high-pitched crisis of acute facet inflammation usually follows a wrenching of the spine which provokes a slower brewing problem. A dormant stiff facet sets itself up for being hurt by not accepting shock as readily as its neighbours above and below—or even its facet partner on the other side.

The cause of the flare-up can be hard to pinpoint but in a matter of hours the injury can be literally crippling. It is usually something awkward but not disastrous, such as moving a pot plant which was heavier than you thought. You often hear a small sound, like a click or a small tear in your back, which may give you fleeting pain at the time but then passes off. But by nightfall or next morning, when the heat of the exertion has gone, a nasty frightening back pain comes on.

At the height of the crisis, the symptoms are a stabbing pain in the side of the back which often goes with searing waves of pain down your leg. Your back feels hard and sore with muscle spasm on one side but the leg pain can be almost unbearable. The pain is often associated with intense pins and needles and a burning sensation which floods downwards as your leg takes weight.

Leg pain of this type is called sciatica, but it is different from the sciatica of a bulging disc (see Chapter 5). Although it is hard for you to know the difference, as a rule, acute facet inflammation gives a hot surging tide of prickling pain down the leg, whereas disc prolapse sciatica is more like intense cramp locking up the leg muscles.

Manual diagnosis of an acutely inflamed facet

Locating the swollen joint with the thumbs usually confirms the facet as the source of trouble. Probing in beside the spine with the thumbs, the inflamed capsule can be felt welling up from below, like the dome of a cathedral rising up through low-lying cloud. It has the typically tense feel of an over-filled hot-water bottle and rebuffs the pressure of probing thumbs.

Apart from the joint feeling hard under the skin, direct pressure usually gives a sudden slice of pain through the back which may then flood down the leg. Unlike this direct evidence of facet involvement, it is impossible to be so sure if the disc is causing sciatica. The disc is around at the front of the spine, out of reach of the hands and, I believe, can only be assumed to be the culprit if there is nothing wrong with the facets. I mention this because there was a time, not too long ago, when all forms of sciatica were deemed to come from herniated or prolapsed discs. I believe discogenic sciatica is infinitely rarer than facet-based sciatica, though much harder to fix.

Typical facet joint sciatica is easily set off by changes in position. Any posture which compresses the swollen joint can exacerbate the pain. Even small positional adjustments can cause a corresponding reaction down the leg, as if the new contortion increases the swelling on the nerve. The pain then fades as the swelling oozes to other parts of the joint capsule and drains away.

In the acute stage of facet arthropathy there is usually armour-plated spasm of the muscles protecting the back. Unfortunately the spasm often makes matters worse by obliterating too much movement and letting the swelling accumulate. At this point, the best course of action is to take anti-inflammatory drugs to reduce the heat, and then muscle relaxants to break the cycle of muscle holding.

This treatment is especially indicated if your spine is listing over, with the hips protruding one side and shoulders the other. This 'windswept' deformity is known as sciatic scoliosis and is caused by the muscles on one side of the spine contracting more than the other. Even though its purpose is to spare the joint, the resulting discord often makes the spine more susceptible to other injury, and makes the current problem harder to fix. At this stage, the best thing to do is take your medication and do gentle knees-rocking exercises in bed to 'milk' the joint.

With severe facet inflammation, the recovery from the acute stage is usually quite rapid—as long as the muscle spasm does not hang on for too long. Fear at this stage is usually the greatest impediment and can slow recovery significantly, sometimes completely. Excessive anxiety or anger (even if it is subconscious) will also lower the pain threshold and create a 'volitional' tension in the muscles on top of the automatic protective one.

All going well, the back is not excessively plagued by anxieties and the muscle spasm releases the joint to get going. As soon as normal movement returns, the joint will be well on the way to recovery; and the sooner the better. Normal movement 'works' the joint properly and disperses the inflammation, and everything assumes normality again. Normality begets normality.

What causes the acute pain?

The facet's response to injury is the same as any other of the body's synovial joints. When a knee or ankle is twisted there is a sudden jerk of pain the moment you do it, and just after it feels wonky but still workable. Within a few hours it becomes more painful as the joint swells. It may reach a peak after a period of inactivity when you suddenly sense the pain stubbornly roosting there and you cannot work it away.

An injured facet joint behaves in exactly the same way. But because of the closer quarters inside the back, with many moving parts of spinal machinery packed cheek by jowl into a tightly confined space, a minor injury can have devastating effects. With so many sensitive structures (not least the spinal nerve) and so little room for anything to swell, a relatively small mishap can cause a crisis.

When a facet joint is wrenched, the synovial lining of the capsule weeps clear fluid into the joint space. It is similar to the way tears well up in the eyes, except the tension of the fluid trapped inside the joint is much more problematic. The engorgement makes the joint semi-rigid from its own bloatedness and it has difficulty sliding and gapping open. The lack of movement makes it less competent at pumping out the fluid, which consequently accumulates even faster. Eventually, pressure from the swelling trapped in the joint causes pain.

As with all joints, special 'mechano-receptors' in the capsular wall are stimulated by the pressure of the fluid, and messages relayed to the brain are interpreted as pain. Pain is also registered by the leakage of inflammatory fluids from the torn tissue of the original joint injury. As their chemical concentration rises, free nerve endings in the joint capsule are stimulated. These are called 'chemo-receptors' and they send off more messages to the brain about pain.

The typical searing pain down the leg is caused by irritation of the nerve root when it becomes embroiled in the joint's inflammation, simply by being so close. As it makes its way past the joint on its way out of the spine it is both physically squashed by the swollen capsule and chemically irritated by the cocktail of toxins coming from its inflamed wall. Things hot up apace when the nerve starts to chafe where it threads past both the capsule and other swollen structures. Eventually it too becomes inflamed and leaks inflammatory fluid.

Interrupting the pain cycle

When I palpate an acute facet problem I cannot feel much at all because the superficial muscle spasm is too unyielding to allow the hands to penetrate. Before proceeding, I induce muscle relaxation by getting the patient to rock the knees to the chest and then roll back and forth over the facet. (Reverse curl-up exercises do it faster if they are not too uncomfortable to do.) The physical movement 'pumps' the joint clear by providing artificial activity to evacuate the fluid. This lowers the tension in the tissues and interrupts the cycle of pain. In the same way that a twisted ankle becomes more comfortable when the swelling subsides and movement ekes back, the pain of an acute facet problem dramatically reduces as the engorged joint empties.

The pain also fades as fresh blood passes through the joint, cleansing away toxins liberated by the damaged tissues. As the stale blood is dispersed it lowers the concentration of metabolites and reduces the potency of the pain messages to the brain.

The chronic phase

Chronic inflammation of a facet joint causes a local patch of pain beside the spine. It typically welcomes the piercing pressure from your own fist or fingers on the spot. Although the joint is several centimetres below the skin it can be felt quite easily, like a brick under a mattress, and you can often elicit a half-relieving sweet pain through your own digging. It usually feels about the size of a squash ball.

Lessening the pain

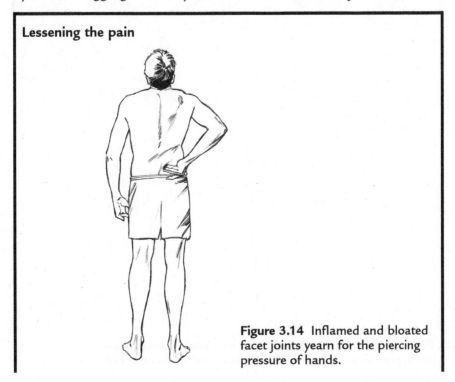

Figure 3.14 Inflamed and bloated facet joints yearn for the piercing pressure of hands.

Depending on the nature of the inflammation, a facet problem may be relieved by stretching or compressing the joint. If there is established tightness and inelasticity of the capsule, you may gain relief by leaning away from the pain and pulling the joint apart. Although the stretching discomfort hurts at the time, the back feels freer and looser afterwards, with less pain. If the problem is more acute, with trapped engorgement in the facet capsule, it can be eased by leaning towards the pain and closing the joint down. This gives a sharper, more piercing pain which can be almost unbearable for a moment but again feels better afterwards. If you arch backwards while still leaning over to the painful side, you can create an even shriller pain, like a knife going in. Alarming as it sounds, compressing the joint like this helps evacuate it and takes some of its bursting discomfort away.

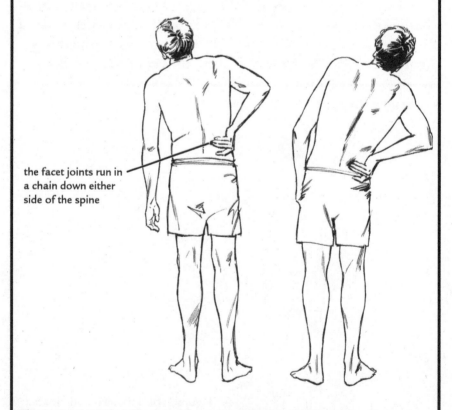

the facet joints run in a chain down either side of the spine

Figure 3.15 'Older' facet problems are eased by stretch, whereas more acute problems are relieved by closing down the joint to evacuate swelling.

You will see in the self-help section that 'milking' an engorged joint is an integral part of treatment. You can use a Ma Roller on the floor to do this (the purchasing details for which are at the back of the book). By lying on your back on the floor and rolling back and forth over the roller with the knees bent, you get the familiar sweet pain as the pressure empties the joint. Take care, however, not to inflame the facet. The Ma Roller is tough medicine when a facet is very irritable and it is easy to stir things up. Do not remain on it more than a few seconds.

It is safer to use a tennis ball when the condition is very severe, because it is safer and kinder on the joint. Its pressure is the nearest thing to the direct contact from human thumbs. But remember it too can make the joint sore and should never be continued for longer than 60 seconds, three times per week.

What causes the chronic pain?

Most of the pain from a chronically inflamed facet joint comes from the stretching of the stiff soft tissues around the joint. As a sequel to disc thinning and overriding of facet surfaces, there is adaptive short-ening of the capsule and the soft tissues which reinforce it.

Mechano-receptors situated in the capsule wall detect the tension in the soft tissues as they are pulled. As small spherical structures between the tissue fibres, they are squashed like tiny ping-pong balls under guitar strings as the tension in the capsular wall mounts. The messages are again relayed to the brain and interpreted as pain.

As an isolated joint becomes tighter and more crimped in the chain it becomes chronically painful. Its capsule is less able to pay out and stretch with the other links in the spine as your body bends and sways about. At a certain point the tightness becomes so unforgiving that every movement elicits a response from the pain receptors embedded in the over-tight fibres.

At this stage, the crimped link is extremely susceptible to addi-tional injury. Its lack of stretch causes fibres of the tight capsule to be tweaked by any chance movement. As each shock racks through the spine, all the mobile segments ride out the force, like a row of corks dancing on the water as a wave passes underneath, except the stiff one which is wrenched anew. Insult is added to injury. Chemo-receptors in the joint capsule are activated by substances released from the torn tissue and their constant bathing of the free nerve endings means the joint emits a barrage of pain signals.

With micro-trauma heaped upon a pre-existing stiffness the familiar old pain becomes a different sort of pain. There is low-grade tenderness in your back and a newer pain in the leg. Pain in the buttock and thigh comes and goes with activities which increase the tension of the facets, such as slumped sitting or prolonged bending activities like gardening. This is called referred pain. The mechanism for this pain is not the same as the direct inflammation of the nerve root which we saw in the acute disorder. Referred pain is a complex phenomenon where structures sharing the same nerve supply as the inflamed joint 'mistakenly' feel pain too. In the same way that the pain of a heart attack is felt in the neck and left arm, nowhere near the heart, the referred pain from an irritable facet joint can be felt quite far away from the point of trouble.

Referred pain rarely extends below the knee although other symptoms can. These can be diffuse, sometimes indefinable sensations which are difficult to put a name to. One leg may not work as well as the other; it is the typical 'gammy' leg and may feel heavier as you walk. The back of the thigh may feel sensitive when you sit, as if the skin and subcutaneous fat is thinner. One foot may feel colder, or as if you have a pebble in your shoe. Sometimes the heel feels numb, or ants seem to be crawling up your calf. Sometimes it feels as if a cobweb is brushing your skin or the leg hairs prickle uncomfortably against your trousers. The buttock of your bad side may feel bonier when you sit, or the hamstring muscle of the problem side tighter. When you bend forward, there may be a tension beside the spine, down through the buttock and into your leg which makes the knee bend automatically as you go over.

Almost all of these signs vary from day to day, sometimes from hour to hour, and almost from one position to the next. They can be explained as the effect of variable swelling within the facet joint capsule. Another explanation points to the build-up of pressure around the nerve root, impinging upon different sensory pathways in the nerve which brings about a wide variety of symptoms.

If there is protective muscle spasm guarding the stiff link, there will be some discomfort coming from the chronically stiffened muscles. This pain is typically a tired cramping feeling, made much worse by emotional tension. Being 'uptight' increases the spinal symptoms by adding to the compression of the spine and congesting of the problem inside.

WHAT YOU CAN DO ABOUT IT

The aims of self-treatment for facet joint arthropathy

With facet joint trouble, the primary objectives are to raise the height of the disc and relieve the congestion of the joint; and then make it run more smoothly. In the acute phase the joint is extremely irritable and treatment is designed to 'milk' it of its swelling. The technique of rocking the knees to the chest does this. Once the joint is emptier and less painful, pivoting on the problem level helps mobilise it. The reverse curl-up exercises work it actively. Abdominal strengthening usually brings about a dramatic reduction in pain because the raised abdominal tone lifts the spine off the painful joint. This exercise also relieves congestion by subjecting the joint to the normal contracting and relaxing of the muscles around it. Briefly using the tennis ball at this stage can also reduce the swelling of the facet joint, but care must be taken because it is easy to make things worse. The correct way to do this is by lying on your back on the floor, knees bent, and positioning the ball under the sore spot. Roll back and forth on the ball, and press the pain away. If a ball is too much then you can use a pair of rolled-up socks. The pressure is less.

In the chronic phase the treatment is much more vigorous. It is aimed at stretching the inelastic joint capsule into its extremes of range. This promotes joint lubrication and cartilage regeneration by alternately closing down and pulling the joint apart. The rapid alterations in pressure contact of the joint surfaces also encourages more active regeneration of the cartilage bed. The Cobra to Child Pose exercise does this. Rolling on a tennis ball or Ma Roller pummels a tight joint capsule and gives the joint more freedom to move.

In the final stages of treatment, disimpaction of the neurocentral core is the ultimate objective. The BackBlock achieves this, and also restores an optimal lumbar hollow. It does this by stretching the front of the hips (the hip flexors) thus reducing an exaggerated lordosis, but it can also correct a pronounced lumbar kyphosis (less commonly a background factor with this condition) by stretching the anterior longitudinal ligament down the front of the spine. While on the Block it is often useful to rotate the pelvis in a minute twisting action which helps disengage the lumbar facets.

Ordinary toe touches are also important. These stretch shrunken facet capsules which, in the case of an exaggerated lordosis, exert a bow-string effect to keep the spine hollowed.

Ultimately, diagonal toe touches pull a tethered nerve root free of adhesions in the small exit-canal (foramen) which may have developed from longstanding inflammation. These exercises also enhance the function of multifidus. The vigour of the action makes the muscles relax properly as the spine goes down and then works them on the way up. Their re-education helps restore joint perfusion (circulation of fluids) and leads to better muscular control in the opening and closing of all lumbar facets.

A typical self-treatment for acute facet joint arthropathy

Purpose:
Ease muscle spasm to relieve compression on the disc, disperse joint inflammation and swelling, strengthen tummy muscles and switch off overactive erector spinae muscles, and gap open the back of the spine to decompress the facet joints and introduce pressure changes to the disc.

Rocking knees to the chest
(60 seconds)
Rest

Reverse curl ups
(five excursions)

REPEAT BOTH EXERCISES 3 TIMES

Rest in bed and use medication under the direction of your doctor. NSAIDs are particularly effective in reducing the pain from the inflamed joint so painkillers are not usually necessary. Repeat exercises every 3 hours throughout the day. When resting, you may be more comfortable in the foetal position with a pillow between the knees to decompress the joint. See Chapter 7 for descriptions of all exercises and the correct way to do them.

For how long? You can progress to the sub-acute regimen when there is no *flooding* pins and needles pain down the leg, either when standing up from sitting or when taking weight on the leg. This can take a week to ten days, though it may be sooner.

A typical self-treatment for sub-acute facet joint arthropathy

Purpose:
Ease muscle spasm to relieve compression on the disc, disperse joint inflammation and swelling, strengthen tummy muscles and switch off overactive erector spinae muscles, break up the brittle jamming of the spinal segments to disengage the jamming of the segments, stretch facet capsules and gap open the back of the spine to decompress the facet joints to introduce pressure changes to the disc.

Rocking knees to the chest
(30 seconds)
Rest (30 seconds)

Rolling along the spine
(30 seconds)

Reverse curl ups
(five excursions)

REPEAT ALL THREE EXERCISES 3 TIMES
Squatting
(once for 30 seconds)

SQUAT THROUGH THE DAY TO EASE PAIN AND TIGHTNESS OF THE LOW BACK

Repeat the regimen morning and evening. After each session, rest on the floor for ten minutes with your lower legs supported on a sofa or soft chair, and a pillow under your head. As a progression of spinal rolling, you can pivot on the painful spot which creates a sharp pain in the back. When up and about, avoid sitting in one position, or standing for too long. (This will usually bring on the pain down the leg.) Have two brief walks per day (less than fifteen minutes). You should walk tall and light, drawing your pelvic floor up and letting your hips swing freely so the low back twists from left to right.

For how long? You should progress to the chronic treatment regimen when there is no leg pain. This usually takes a week to ten days but it may be much sooner.

A typical self-treatment for chronic facet joint arthropathy

Purpose:
Ease muscle spasm to relieve compression on the disc, disperse joint inflammation and swelling, strengthen tummy muscles and switch off overactive erector spinae muscles, gap open the back of the spine to decompress the facet joints to introduce pressure changes to the disc, break up the brittle jamming of the spinal segments to disengage the jamming of the segments, encourage cartilage regeneration of facet joints, stretch facet capsules, strengthen segmental control of multifidus, decompress spine to promote disc repair, and release nerve root tethering.

Rocking knees to the chest
(60 seconds)
Rest (30 seconds)

Rolling along the spine
(60 seconds & also pivot
on the painful spot)

**Cobra to Child Pose
and squat**
(back and forth 3 times)

Segmental bridging
(up and down 3 times)

BackBlock routine Step 1   (60 secs)

Step 2   (30 secs)   Step 3   (15 times)

REPEAT ALL FIVE EXERCISES 3 TIMES

Repeat program every evening, and continue with NSAIDs (anti-inflammatories). At this stage you can often feel the soreness of the joint in the side of your back and the muscles clenching if it gets too congested.

For how long? Continue indefinitely.

When you no longer have *leg pain* you can include the following two exercises in your routine:

Floor twists
(three to bad side, one to good)

Diagonal toe touches
(down to each foot 3 times)

4

The acute locked back

An acute locked back occurs when the spine slips slightly askew at one of the facets.

WHAT IS AN ACUTE LOCKED BACK?

An acute locking episode is when an unguarded movement causes an agonising jolt of pain like a high voltage current to shoot through your back. The pain always strikes at the beginning of a movement, like a bolt from the blue, and leaves you bent over rigid and unable to straighten.

Usually the pain catches you so badly you cannot stir. You cannot go forwards or back, or put one foot after the other. It is a real crisis. Often it makes the knees buckle so you collapse to the floor and an injection of pethidine may be needed before you can be moved. It is always a very frightening experience, and often clearly remembered many years later.

There are an infinite number of minor ways you can suffer an attack like this. You can do it turning over in bed, getting out of a car, pulling your chair out, bending forward to pick up a toothbrush, lifting a bale of wool. One of my patients did it zipping up a ball dress! Common to all is the unexpectedness and a certain lack of exertion. In fact, absence of effort and preparedness for what you were about to do, seem to play a key part.

As a therapist, I find acute facet locking one of the most daunting

Figure 4.1 A facet joint locking can make it impossible to move.

conditions to treat. At the time of the crisis patients are in extremis; they are loath to move and almost hysterically fearful of doing anything that might cause another jolt of pain. Long after the original episode has passed they remain fearful of it happening again and often feel their back is never the same again (indeed some feel their *life* is never the same again).

Over the years, many opinions have been put forward as to what goes wrong. Although acute locked back feels as if something has slipped out, it is most definitely *not* a 'slipped' disc. However there is never any objective evidence to explain exactly what has happened. There is nothing to see on the X-rays or other forms of imaging, and the neurological assessment is usually clear. Yet there is a fellow human cast down and immobilised, often on the floor as if caught in a freeze frame, literally rigid with pain.

One popular explanation is the jamming of a meniscoid inclusion (a tiny wedge of cartilage from the margins of the facet joint) between the two joint surfaces, which sends all the muscles of the spine into a gust of protective spasm. A similar, more plausible, explanation

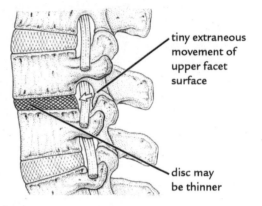

tiny extraneous
movement of
upper facet
surface

disc may
be thinner

Figure 4.2 Sometimes an unguarded movement without the tummy muscles adequately braced can allow one of the facets to slip slightly askew in its joint.

points to the nipping of the sensitive folds of synovial membrane between the facet joint surfaces.

I believe the main cause of facet locking is a momentary lapse of spinal coordination, causing one of the facets at the back of the spine to slip slightly askew in its joint. Almost before a movement has begun, the spine is caught unawares and the facet moves slightly out of alignment. If pinching of the synovial lining does happen, I suspect it is only part of the wider picture of the facet 'mis-jointing' and then jamming.

The degree of movement is infinitesimal so it is never possible to take a picture and see any joint dislocation. But the actual slip is not the problem. The reaction is: an instantaneous and massive protective response from the muscles which takes your breath away as they lock up the spine.

A joint being 'out' anywhere else in the body does not arouse the same defence. (We have all experienced a wonky knee when our kneecap temporarily mistracks.) However, the spine's heavy responsibilities for structural support and protection of its fragile festoonery of nerves inside make it keenly alert to any threat to its integrity.

When facet locking happens in the neck it is relatively easy to manipulate back into position. The neck's slender accessibility makes it much easier than the low back where bulky protective spasm quickly sets in, making it difficult to pull the segments apart. If you are lucky and get to an osteopath/chiropractor/physiotherapist soon enough, a small high-speed manipulative thrust, with its characteristic popping sound, can break the jamming of the joint and realign the vertebra correctly. ·

These are the wonder cures you occasionally hear of. The technique momentarily gaps the joint open and lets it slot back together in proper apposition. If it is successful, the joint immediately rides freely again, and you can walk away on air with none of your former pain. Any residual muscle spasm will be gone within a day or so.

More usually though, by the time you get to a practitioner the stiffness of the muscles is so well established it stops the joint being physically opened. Attempting manipulation at this stage only makes matters worse. It alarms the patient (because it hurts) and causes the protective hold of the muscles to intensify.

CAUSES OF AN ACUTE LOCKED BACK

- A natural 'window of weakness' early in a bend
- Segmental stiffness predisposes to facet locking
- Muscle weakness contributes to facet locking

A natural 'window of weakness' early in a bend

All spines experience a natural vulnerability on bending until they get themselves properly braced. I believe that facet locking occurs when the spine is caught momentarily ill-prepared as it passes through a fleeting 'window of weakness' in the early part of range.

Bracing happens when the muscles at the back and front of the abdomen contract in unison to stiffen the spine. They create a valuable tensile strength which keeps the spinal segments secure until they can be passed into the care of the strong system of muscles and ligaments running down the back of the spine. These then pay out slowly and lower the spine forward like a mechanical crane. However, the powerful erector spinae muscles and the 'posterior ligamentous lock' do not come into their own until the spine is well forward into a hoop when at last they generate sufficient tension to make the spine safe.

Up until this time the spine passes through an un-sprung phase when it must rely on the tummy muscles to tense the abdomen and slightly hump the spine to get it across the wobbly part. This slight tensing and rounding of the lower back in preparation for bending plays a subtle but invaluable role in putting the important multifidus and transversus abdominus muscles at a better angle of pull. They control the tipping of the segment by being intimately involved in the gapping apart of the facets to allow the segments to go forward.

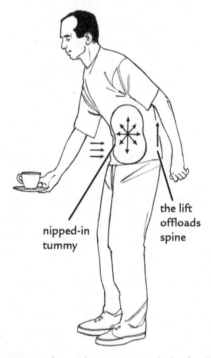

nipped-in
tummy

the lift
offloads
spine

Figure 4.3 It is important to have the tummy switched on to keep the spine braced and the segments stable through the 'window of weakness' in the early part of bending.

But even the slightest delay of one or other partner in the co-contraction can cause a glitch in the movement. When the spine starts to move before both systems are ready, it is caught off-guard and minutely disjoints somewhere in the column at one of its facets. The threat this poses to the spine causes a massive protective response from the muscles which jams the slipping facet before it can go any further. This is the reaction which brings you to your knees.

A locking incident often happens when people are recovering from a viral illness. Generalised debility is the most likely explanation, when the reflexes are dulled and the tummy muscles (in particular) cannot generate a quick enough response to keep the spine braced.

Facet locking can also happen a day or so after some form of serious exertion such as laying paving stones, cutting timber or digging in the garden. In these circumstances, it is probably the overactivity of the long back muscles and their residual raised tone which disturbs the natural harmony of the two deeper groups working underneath. The story is always the same: your back has been feeling stiff for a day or so when it was harder than usual to keep the tummy pulled in. Then some minor incident—almost too incidental to be taken seriously—catches unexpectedly and brings you down.

Segmental stiffness predisposes to facet locking

Segmental stiffness, even when benign and painless, can predispose your back to locking if the intervertebral disc between the two segments has already lost its oomph.

One of the specific roles of the multifidus muscle (with its willing helper the 'muscular' ligamentum flavum on the other side of the facet) is to pre-tense the intervertebral disc at each lumbar level. As soon as your spine starts to move, the disc should be as tense and plump as possible to prevent any wobble of the vertebra. If the disc has already lost fluid and the intradiscal pressure has dropped, it is much harder for the facet muscles to get the disc primed. Thus a spine already in line to develop symptoms from a stiff segment is also more likely to suffer a facet locking incident.

If the disc has already dropped in height and the other ligaments holding the segment in place have become slack, the segment is additionally vulnerable. The bone-against-bone locking of the facets (which provides a more basic tier of stability) is not specific enough to prevent minute movement of a vertebra and the facet can slip slightly askew unless the volitional muscular hold (the tummy) is compensating well. In fluke circumstances, an ill-prepared tummy can bring the whole lot down.

Muscle weakness contributes to facet locking

Longstanding segmental stiffness often makes the muscles weak. When a segment is too stiff to be active the small muscles which work it have less to do, which makes them atrophy. This particularly applies to the deeper fibres of multifidus which lie right on the back of the facets, acting as their special protector.

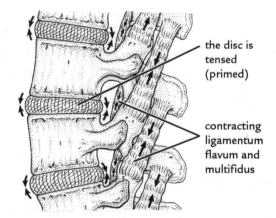

the disc is tensed (primed)

contracting ligamentum flavum and multifidus

Figure 4.4 Contraction of the two muscles closest to the disc (multifidus and ligamentum flavum) clamps the segment which 'primes' the disc and prevents wobble during early bend.

In its moment of need, multifidus may find when its segment goes to slip it is not up to the task of keeping its joint under control. This is particularly important when a back has been suffering underlying trouble for some time. When there is low-grade inflammation of a facet, there is evidence to suggest that multifidus 'deliberately' under-performs to spare the irritable joint excessive compression. Although this might be for the prevailing comfort of the inflamed facet in the short term, in the long term it leaves it without the muscular control to compensate for other inefficiencies. A problem facet is wide open to a future locking episode.

Apart from this automatic inhibition, primary weakness of muscles can also cause facet locking. General indolence or unfitness can so impair our streamlined coordination that the tummy and back muscles fail to cooperate simultaneously in holding the spine supported. In a brief blip of control they perform out of sync, also making it difficult for the deeper intrinsics to come in at just the right time. If they fail to hump the back in its imperceptible first few degrees of bending, and the two important deep muscles cannot develop an optimal line of pull, the fundamental unit at the centre of the motion segment—the disc—does not get primed properly, and the segment can slip.

As a first cause, a weak tummy is easy to blame because it affects so many of us. Most of us have weakness (to varying degrees) of both the abdominal wall and pelvic floor. Weakness here creates a sloppy hydraulic sack which is less efficient at forcing the spine skywards. With a weaker up-thrusting force within the abdominal cavity and failure to round the back, there is a reduced tensile strength between the spinal links, leaving them more susceptible to jostling about when the spine goes to move.

Women are particularly susceptible to this in late pregnancy and early motherhood. When the muscles are stretched and weak and the ligaments still soft from the pregnancy hormones, it is easy to be less prepared for spinal action. This can also happen to any of us suffering from exhaustion, lack of fitness or recent weight gain. Getting up after an illness is also a time for suffering an acute locking incident, probably because of generalised weakness. Food poisoning and the flu are also commonly mentioned as predisposing factors.

The relative weakness of multifidus in its role of preventing sideways twist of a vertebra may also contribute to facet locking. Since all functional bending movements incorporate a twisting component (unlike robots which have an up–down and left–right action only) multifidus is like David versus Goliath in steadying the torque of the heavy trunk above. Even though the available range of segmental rotation is a few degrees only, multifidus (acting one-sidedly) is the only muscle which directly controls its vertebra. It does so at the very beginning of range by hanging on to its tail and preventing it from going both forward and twisting sideways. (Iliocostalis, another intrinsic muscle, also controls the vertebra twisting but only deeper into the forward bend.) All the other muscles controlling spinal rotation are in large sheets on the surface of the trunk with no direct attachment to the spine.

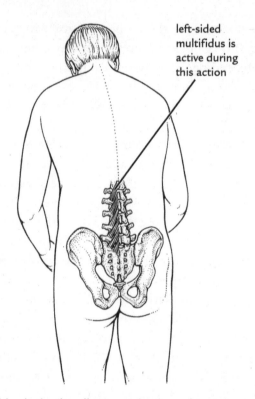

left-sided multifidus is active during this action

Figure 4.5 Multifidus is the 'bending' muscle. As we bend down to the left, the left-sided multifidus hangs on to the tails of the vertebrae and steadies them from swinging to the right.

THE WAY THIS BACK BEHAVES

The acute phase

The electric jolt of pain comes at the beginning of a movement—almost before it has started. In a split second there is an ominous sort of 'uh ohhhh' feeling, as if your spine is about to do something it shouldn't. The action is usually inconsequential—you can be leaning forward to pick up a coffee cup and your whole world stops. Apart from the suddenness, you are incredulous that something so trivial could have brought you so undone.

The sudden clench of pain completely takes your strength away. You clutch furniture for support and then, with your hands sliding down your thighs, you might slither helplessly to the floor. There at least you are more comfortable but you are like a beached whale and cannot be moved. If you are alone when it happens, it can take hours to crawl to the telephone to call for help.

The pain at this stage can alternate between a cramp hovering in the background and excruciating jolts whenever you try to move. If you need to move a leg you have to inch it across using a sideways heel–toe action on the floor. If you attempt to lift the leg or jerk in any way the pain will zap you again and leave you gasping.

What causes the acute pain?

The grabs of pain in the acute phase come from the muscles locking up the whole spine to trap the individual joint. They jump instantaneously into a high-pitched clench whenever they sense the joint going to move. The muscle contraction stops the mini-dislocation going any further but it also prevents the joint disengaging and repositioning correctly. The muscles keep on keeping on, like a dog with a bone, and they are a major part of the problem.

The intense compression of the joint while it is still out of kilter sends out the usual alarm signals of any traumatically twisted joint. The back does not let you off as lightly as a twisted ankle, probably because of the complexity of the workings inside and the relative size of the tiny joint compared to the bulk of the muscles guarding the spine. Until they are satisfied they can relax, they stay guarding the joint, keeping it out of action and locked away in the machinery of the spine.

The special mechano-receptors in the capsule let the brain know the joint is locked under pressure. They do so the instant the joint freezes and repeat the message every time there is even so much as a flicker from the muscles. A different sort of lower grade pain creeps in several hours later from stimulation of the chemical receptors in the joint capsule. They register the build-up of toxins in the tissues, both from the original capsule-wrenching damage and the stagnation of circulation through the capsule. As the concentration of toxins rises, the protective spasm increases, which intensifies the hold on the joint, and the pain coming from it.

The muscle spasm itself can cause a similar type of residual pain. When blood must be squeezed through tonically contracted fibres, the metabolic waste products cannot get away. As their concentration rises, their irritation of the free nerve endings in the joint's tissues is read as pain. Cramping muscles also experience another type of pain from lack of oxygen (anoxia). The over-working muscles cannot get sufficient fresh supplies through, creating a typical tired pain with peaking pin-prick twinges.

Pain begets more spasm which begets more pain and the cycle intensifies unless you get the joint moving. For this reason, reducing the muscle spasm and restoring activity is very important early in the treatment regimen.

When you have just been struck down however, any sort of therapy seems a long way off. At the time, the pain seems to come from everywhere and the back feels frighteningly locked.

At this stage, the best course of action is an intra-muscular injection of pethidine (a strong painkiller) and a muscle relaxant such as Valium as well. The first priority is to get you off the floor and into bed and the quicker a doctor is called the better. For your future rehabilitation you need to get over the incapacity stage as soon as possible, almost as much for your head as your back.

If the first attack is not handled properly you may never get over it, physically, mentally or emotionally. Many people with ongoing troubles claim their problem started with an incident like this which was never properly resolved. Twenty or thirty years later they can remember every detail and let you know that their back has never been right since.

The sub-acute phase

Within a matter of a few days, the crisis of the acute condition should pass. With resting in bed and proper medication the muscle spasm relaxes and it is easier to move. Your own attitude makes a big difference here. The more fearful and tense you are the more you hold things up. Breathing quietly, keeping calm and deliberately making your spine move again helps break through the physical and mental barriers and relieves the pressure on the jammed joint. The more anxious you are, the slower this resolution is.

As the muscles relax, it becomes easier to lift your bottom off the bed although it is still painful to turn over. Slowly the guarding reaction loses intensity and the back softens its over-vigilant armour-plated hold. There are no crippling jolts of pain if you move slowly. Unless you make a sudden jerking movement or sneeze or cough you will be able to get up, although it is difficult doing something complicated like getting out of bed.

Slowly the broad expanse of pain retracts to a localised area of soreness and it is easier to pinpoint the focus of the trouble. By this stage your back usually feels bruised and fragile, as if it has been through an ordeal. Even though it is weak, it is ready to get moving.

The chronic phase

In its chronic phase this problem behaves the same as facet joint arthropathy (see Chapter 3). When the blanket of protective spasm lifts, the dysfunctional joint underneath emerges through the mist. It needs to be mobilised as soon as possible and brought up to par with the rest of the joints, otherwise the problem becomes chronic and continues off and on indefinitely.

When the damaged facet is slow getting going the protective muscle spasm hangs around and the condition worsens. There is shrinkage of the joint capsule as a legacy of the scar tissue formation but, in a seemingly contradictory way, the capsule may also be left weak. Microscopic scarring cobbles the joint and pinches it tight, which leaves it stiff, but the original renting of the joint capsule and the weakness of the local muscles around it leave it vulnerable and easier to re-injure.

Taken to its extreme, the facet joint may eventually become un-stable (see Chapter 6). This condition brings with it a conundrum for the joint's management. How do you strengthen the stiff, inelastic joint capsule when its very stiffness may be the only thing holding it together? This is the problem facing all facet instabilities, and it is not an easy one to deal with. Better therefore to handle it early on—after the first facet locking episode—so you never have to deal with the difficult end.

The aim is to get the joint going early to lessen the scarring. Even if your problem is longstanding (when you fear that loosening the joint will allow it to it slip again), the joint must, nevertheless, be mobilised, while making sure to cover the new-found freedom with improved power of the segmental muscles (mainly multifidus). The most effective way of doing this is by intrinsic exercises, unfurling the trunk off the end of a table, but an earlier and easier version (though with shorter leverage and less empowering) is simply bending forwards to touch your toes and uncurling cog by cog up to vertical.

If the intrinsic power of the segment is not restored quickly you are left with a back you keep hurting with twisting movement. You bend down to help an elderly lady with her shopping and you feel the familiar tweak as you overtax the weak facet. By next day your back has stiffened and developed its familiar lateral 'S' bend with one hip protruding. It feels tighter and caught up on one side and you keep digging your fingers in to find relief.

People often seek treatment at this point because they find they can do progressively less before they tweak the facet again, with it taking longer each time to recover. Whereas it used to be two or three days in bed now it takes ten and you are barely over one attack before the next one comes along. One episode seems to merge with the next.

WHAT YOU CAN DO ABOUT IT

The aims of self-treatment for an acute locked back

The immediate aim in treating an acute locked back is to quell the alarm, so at the very least you can move your limbs without pain and turn over in bed. After the crisis has passed, it is important to deal with the joint sprain, and then bind up the problem joint with good muscle support so it does not happen time and again.

Getting the muscle spasm to relax is best achieved by muscle relaxant drugs and strong painkillers, both administered by injection. As soon as the drugs start to work, the spine must be exercised to lessen the clench of the muscles and release the joint. This is started as soon as possible—gathering one leg at a time up to the chest and rocking the knees, hand cupped over each knee, to make infinitely small and coaxing oscillations. Remember, it *is not* a vigorous tugging action. Continue for as long as possible (several minutes), gradually feeling the hardness in the back melting and the movement getting just that much easier. This gentle rocking can be repeated innumerable times throughout the day. It usually takes less than 24 hours to be able to move your legs in bed comfortably and to sit up without difficulty.

The sooner the next phase can be commenced, the faster the problem resolves. Both relaxation of muscle spasm and the return of normal movement of the injured joint are achieved by gentle reverse curl-up exercises. Working the tummy muscles hard relaxes the spasm of the long back muscles and encourages normal hinging of the locked vertebra. As soon as the joint starts moving, the trapped engorgement disperses and the pain eases dramatically. In many respects, the treatment at this stage is similar to the chronic phase of facet arthropathy, although there is a greater emphasis on re-educating the muscles to control the wrenched facet.

The final stage of treatment is devoted almost entirely to improving both the strength and the coordination of the various muscles influencing the injured joint. The strength of the deep muscles compensates for the traumatic stretching of the capsule and ensures the joint is not left susceptible to repeated lockings. At the same time, stretching of the long erector spinae muscles, particularly into full bending movements, inhibits their tendency to overact which automatically keeps the deeper ones weak. The bending exercises (toe touches) also relieve the general stiffness of the back.

A typical self-treatment for acute locked back

Purpose:
Ease muscle spasm to relieve compression on the disc, disperse joint inflammation and swelling, strengthen tummy and switch off overactive erector spinae muscles, and gap open the back of the spine to decompress the facet joints and introduce pressure changes to the disc.

Rocking knees to the chest
(60 seconds)
Rest, legs crooked, knees propped
together (30 seconds)

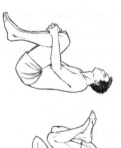

Reverse curl ups
(five times)
Rest

REPEAT BOTH EXERCISES 3 TIMES, TAKING CARE WHEN
RETURNING EACH LEG TO THE BED AFTERWARDS

Medication: intramuscular pethidine followed by muscle relaxants
and anti-inflammatory (NSAIDs) tablets administered only by your
doctor. Rest in bed. Repeat treatment regimen every 2 to 3 hours
through the day or less often if you are sleepy from the drugs. When
commencing the rocking exercise, always lift one leg to your chest at
a time, drawing in your tummy and pelvic floor. The legs are very
heavy and if you try hauling both up together you will jerk your back
and set off another seizure of pain. See Chapter 7 for description of all
exercises and the correct way to do them.

For how long? If you can begin the medication soon enough while
doing the exercises for acute locked back, you may be able to start
the sub-acute treatment by the next morning.

A typical self-treatment for sub-acute locked back

Purpose:
Ease muscle spasm to relieve compression on the disc, disperse joint
inflammation and swelling, strengthen tummy and switch off over-
active erector spinae muscles, gap open the back of the spine to
decompress the facet joints and introduce pressure changes to the
disc, and strengthen transversus abdominus to provide core stability.

Rocking knees to the chest
(60 seconds)
Rest (30 seconds)

Rolling along the spine
(60 seconds)

Reverse curl ups
(five excursions)

Legs passing
(five times each leg)

REPEAT ALL FOUR EXERCISES 3 TIMES

Rest after each session with your lower legs supported on pillows. Repeat every 3 hours but do not hurry. Always expect the first one or two reverse curl ups to be more painful and try to round the lower back.

For how long? You must stay on this sub-acute regimen until you are largely pain free doing the legs passing and there are no sudden seizures of pain with unguarded movement. This usually takes two to three days to achieve.

A typical self-treatment for chronic locked back

Purpose:
Ease muscle spasm to relieve compression on the disc, mobilise central core and facet joint which disperse joint inflammation and swelling, strengthen tummy and switch off overactive erector spinae muscles, gap open the back of the spine to decompress the facet joints and introduce pressure changes to the disc, strengthen transversus abdominus to provide core stability, strengthen tummy, stimulate facet joint cartilage regeneration, decompress the spine, strengthen multifidus muscle to control the facet, strengthen multifidus one-sidedly to re-establish intrinsic muscles' segmental control of the facet joint, and release nerve root tethering.

Rocking knees to the chest
(60 seconds)
Rest (30 seconds)

Rolling along the spine
(60 seconds & also pivot
on the painful spot)

Legs passing
(five times each leg)

**Cobra to Child Pose
and squat**
(back and forth 3 times)

BackBlock routine Step 1   (60 secs)

Step 2   (30 secs)   Step 3   (15 times)

REPEAT ALL FIVE EXERCISES 3 TIMES

Regimen repeated every evening. Your back will feel generally sore
and fragile when caught off-guard but there is no sense it will let you
down and you will enjoy the stretch of normal movement returning.
It will feel tired and achey if you have been on your feet for too long.
When this happens you should lie down and gently rock your knees
to your chest until the pain goes.

**When you no longer have *leg pain* you can include the following
three exercises in your daily routine:**

Toe touches
(down and unfurling up to vertical,
twice)

Diagonal toe touches
(down and up to each foot 3 times)

Spinal intrinsics strengthening
(5–10 excursions, *every ten days only*)

For how long? Continue this regimen indefinitely.

The prolapsed 'slipped' intervertebral disc

A prolapsed or 'slipped' disc occurs when the vitality of the disc breaks down and the back wall bulges where it perishes.

WHAT IS A PROLAPSED DISC?

A 'slipped' disc is a localised bulge in the back wall of a disc. It is caused by the devitalised nucleus losing cohesion and trying to escape through a chink in the cracking disc wall (anulus). The condition needs explaining because for years it has taken the blame as the main perpetrator of back mischief. In the 1930s, 'slipped discs' were nominated as the chief cause of back trouble, and this limited view of what goes wrong has held sway, almost to the present day. It was thought that a disc could pop out of spinal alignment—like a saucer slipping sideways out of a stack—and pinch a nearby spinal nerve. Sometimes a disc wall does buckle (or prolapse) and this painful distension does not disappear when segmental compression is alleviated, but by modern estimates fewer than 5 per cent of back problems are caused in this way.

True prolapse is a localised bulging of the disc wall, caused when a nucleus is 'rendered expressible' by the degenerative process. Inordinate compression (of the type caused by heavy lifting or when the segment remains locked over a period of time by intense muscle

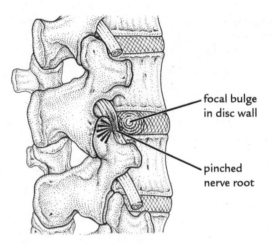

focal bulge
in disc wall

pinched
nerve root

Figure 5.1 Disc prolapse is a localised bulge in the disc wall caused when the degenerated nucleus loses osmotic cohesion and tries to escape through a crack in the disc wall.

spasm) can make the nucleus track out of the disc's centre by burrowing through a small chink or radial fissure in the wall. The escaping pocket of nucleus creates pressure against the sensitive outer 'skin' of the disc. The painful stretching of the intact outer layers of the wall creates a particularly unpleasant cramping ache, deep within the back.

The localised mass of a prolapsing disc can also annoy other pain-sensitive structures nearby. A 'posterior bulge' directly out the back of the disc can stretch the highly sensitised posterior longitudinal ligament and also obtrude into the bundle of nerves hanging down inside the spinal canal (the cauda equina). This can cause disturbed bladder control and/or saddle anaesthesia. On the other hand, a postero-lateral prolapse comes out the diagonal back corner of a disc and may squash and stretch (at the same time) the spinal nerve root nearby in its exit canal (foramen). This causes severe leg pain (sciatica) which can be sharp and 'lancinating' in its acute phase or a deep aching cramp when more subdued. The pain is often accompanied by numbness, weakness, positive tension signs in the nerve, and loss of reflexes.

Sometimes the wayward nucleus ruptures right through the few remaining layers of a painfully stretched anulus. This is known as fenestration (or sequestration) and accompanies a kind of crisis with

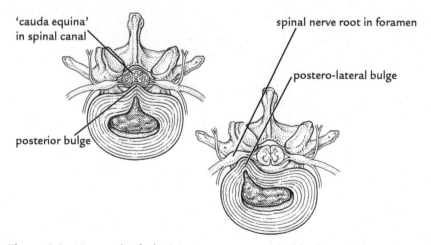

Figure 5.2 A 'posterior bulge' may annoy nerves inside the spinal canal whereas a 'postero-lateral' bulge, in either back corner, may impinge on the nerve root in the exit canal (foramen).

the back, usually as a consequence of a forced lifting movement. At this point, the back pain abruptly disappears, with the release of pent-up pressure behind the wall. However, the leg pain dramatically worsens as the nerve root—though released from the mechanical stretch—is now chemically inflamed by the highly toxic nuclear material. Drastic as it sounds, the extruded nucleus is quite quickly absorbed by the blood stream and surgical removal is not required. Usually, within a couple of weeks, there is very little pain and the problem reverts to a grumbling low-grade discomfort more typical of a stiff spinal segment.

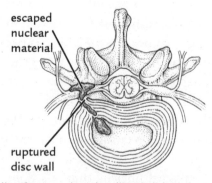

Figure 5.3 With disc fenestration the disc wall bursts and the toxic, semi-degraded nucleus is disgorged, adding further irritation to the already sensitised nerve root.

When disc prolapse occurs it is always part of the wider picture of segmental breakdown, when the degenerative process has rendered the nucleus expressible. Neither disc prolapse nor fenestration happens when a spinal segment is *healthy*. This has been demonstrated in the laboratory when spines of young cadavers were progressively loaded until the bones crumbled rather than the discs rupturing. Even cutting the wall of a healthy disc with a scalpel, and then subjecting it to load, will not cause the nucleus to prolapse or rupture, although this does initiate the disc's slow breakdown in the longer term, as the wall is attacked by enzymatic action.

A disc never slips to cripple you with one blinding movement. Healthy discs are extraordinarily strong and never dislodge with one ill-considered move. Discs are very superior segmental connectors. They make powerfully bendable fibrous unions between the vertebrae and are one of the main agents keeping the spinal segments together. Believe me, nothing slips freely anywhere.

Perhaps the pure expressiveness of the term 'slipped' has captured the imagination of patients and specialists alike and slowed the pace of change. When a back is devastatingly and persistently brought low, the very word conjures images of something bad; of something slipping off centre and jamming the works, even though spinal mechanics are far too sophisticated for anything so crude to occur. It is ironic that for so long so much has been attributed to such an unlikely cause; that this rare condition could have obtained such wide currency as to be inversely proportional to its true incidence.

Mild disc prolapse is extremely common and usually asymptomatic. This has been borne out with the advent of 'magnetic resonance imaging' (MRI) where large cross-sections of populations *not suffering back pain* can be scanned, minus the radiation hazard of X-ray myelography. Results have shown an astonishing one in five under the age of 60 and one in three over the age of 60 with disc prolapse—all without symptoms. Nearly 80 per cent of those scanned had bulges.

Healthy discs broaden by half a millimetre or two as they take weight and all healthy discs bulge considerably more by the end of the day when they have lost up to 20 per cent of their fluid. Less healthy discs bulge more readily because they have a lower proteoglycans content and consequently have trouble holding their fluid. Like a car tyre with a slow leak, degenerating discs are drier, so they have a low hydrostatic pressure which causes their walls to bulge.

But there is another factor which comes into play once a spinal segment has started degenerating: muscle spasm. Any pathology of a motion segment causes protective spasm of the spine's extensor muscles. If the protective response is intense and continuous, eventually the compression makes disc distension irrevocable as enzymatic action attacks the walls. Though impossible to say how many Newtons of compression force is exerted by the clenching hold of muscle spasm, the sustained loading causes disc cells to die. In a direct analogy to the rubber of a car tyre perishing, a buckling wall of the disc cannot be 're-inflated' when it has been flat for too long.

This makes it only too clear how urgent spinal decompression and muscle relaxation exercises are when backs are in extremis. When a disc has remained buckled for longer than a few weeks it is no easy job to rehydrate it. This factor alone has implications for the urgency of effective spinal treatment when a back problem is in its acute stage. Even if the original pathology is not discogenic, the disc will eventually be destroyed if it remains compressed by muscle spasm. These are the cases which require surgical excision of the distending disc wall; formerly laminectomy and more latterly, discectomy.

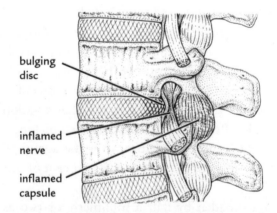

bulging disc

inflamed nerve

inflamed capsule

Figure 5.4 When a spinal segment is locked by the compression of muscle spasm all its soft-tissue structures inflame and swell. In this instance, the disc is the easiest to section out, though its removal makes the facet bear more load and its capsule pucker.

Disc prolapse is really a culmination of events; a secondary or tertiary phenomenon rather than a primary entity. True disc prolapse always has a history of pre-existing breakdown of the segment (see Chapter 2) usually with a grumbling backache over several years, though it may have been fairly 'silent' with few symptoms along the way. Eventually it becomes something different, as the original back pain disappears and a new type of pain develops in the leg.

Diagnostic techniques

In the past, discs were difficult to 'see'. Because disc material is not radio-opaque it is not possible to get a clear picture on X-ray. To find out if a disc was protruding into the spinal cord (in the spinal canal) or the spinal nerve (in the intervertebral exit canal), a radio-opaque dye was injected into the space around the cord and its investments, then the patient tilted this way and that to make it trickle around the discs. Afterwards an X-ray was taken to reveal the discs' outline. The procedure is called a myelogram.

Fortunately this very clumsy and unpleasant procedure (which often left the patient with headaches for days but much more seriously, sometimes caused 'arachnoiditis'—a longstanding inflammation of the tissue coverings of the spinal cord) has been completely superseded, first by CT scans and then by MRI (magnetic resonance imaging). Although MRI in particular is expensive, it shows increasingly clear (almost 3D) images of soft tissues and bone alike, making it much easier to interpret the state of play of all spinal structures, not just the discs.

It is impossible for people like me to feel the discs with our hands because they are around the front of the spine and far out of reach of our probings. It is only possible to feel the state of the neurocentral core by palpating it through the vertebra's backward tail (the spinous process). Although there may be a typically 'gummy' feel of the surrounding tissues when there is a tense disc bulge, this may be difficult to pick up. Sometimes however, the slight palpatory pressures disturb a bulge and provoke pain further afield, probably by the swollen wall rubbing against the nerve root.

If minimal pressure brings on a cramping leg pain, it spells out a high irritability of the nerve—although it is important to exclude facet involvement, rather than disc. This is done by palpating 1–2 cm laterally from the centre, over the facets.

Because it is impossible to feel the disc we must rely on objective signs to show if the spinal nerve root is under duress. They are all called 'neurological signs' and they indicate the degree of irritability of the nerve and the extent to which it has stopped working. The straight leg raise test (SLR) involves raising the leg to a right angle and not allowing the knee to bend. By increasing the tension on the nerve roots you can tell if one is inflamed. If there is a marked nerve inflammation the leg barely gets off the bed before it exacerbates the leg pain. Other signs of nerve involvement are dulled or absent reflexes (behind the ankle and below the knee), numbness of the skin on the leg and loss of muscle power. It is confusing that severe inflammation of a facet joint gives almost the same signs (see Chapter 3). I believe you can only conclude that symptoms are caused by disc prolapse if there are disturbances of bowel and bladder function (which a facet cannot cause).

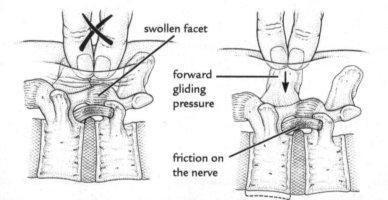

swollen facet

forward
gliding
pressure

friction on
the nerve

Figure 5.5 Thumbs can feel through to a swollen facet but not the disc. Indirect pressure delivered via the spinous process can provoke leg pain by causing friction of the disc bulge against the nerve.

True disc prolapse is one of the most difficult spinal conditions to fix conservatively but, given time, it will eventually regenerate. Even with stark evidence of disc prolapse on MRI it can be nothing short of miraculous how well cases like this can do. Admittedly, once the nucleus is out of the centre of the disc it is difficult to get back (it is often described as getting toothpaste back into a tube) but mobilising the segment releases it from compression which *improves disc nutrition.* The manual loosening also relaxes the muscle hold which lessens the pressure on the bulge and allows the disc to imbibe fluid. It also frees the disc to suck and blow to promote cellular activity and repair. And finally, the introduced movement also restores a better circulation of blood through the whole area (not the disc which is avascular) to reduce inflammation caused by the many other swollen structures of the segment, all pressing and chafing against one another, the disc being but one.

Coaxing a crimped spinal segment to move can dispel even the most menacing leg pain, even though a nerve root, once inflamed, retains a lowered threshold for some time. For many months it may be susceptible to flare-ups, especially after periods of slumped sitting, which stretches the covering (dura) of both the spinal cord and the nerve roots. With even the vaguest return of muscle spasm, or if the disc's circulation becomes sluggish for other reasons, shades of the familiar leg cramp can start up.

Disc surgery

The fact that the whole 'metabolic climate' inside an inflamed segment contributes to the irritation of a nerve root may explain why removing a disc with surgery is so often unsuccessful. Some figures estimate that 50 per cent of operations for a slipped disc leave the patient worse or at least no better. Removing the disc may not be removing the problem; it may be worsening it. If indeed the facet is the main source of pain, wholesale disc removal obliterates the disc space and brings more pressure to bear on the facets. After the operation the leg pain is much worse—which is very disappointing after all you have been through. No sooner are you upright than all your symptoms return, as bad as they ever were.

However, many back operations are extremely successful. In the past, a more radical procedure called a laminectomy was performed which removed the entire disc (like tearing away three-dimensional pieces of fingernail with pliers) and then part of the bony foramen

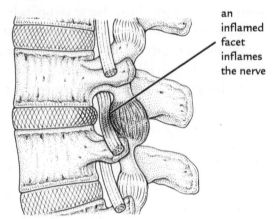

an
inflamed
facet
inflames
the nerve

Figure 5.6 If both disc and facet capsule are swollen, changeable leg symptoms are more likely to come from the facet. The facet's rich blood supply makes its behaviour mercurial.

above and below the nerve root was removed with bone cutters. Sometimes a spinal fusion would then be performed during the same operation to deal with the instability created by destroying the fibrous union. A fusion would be performed either by packing out the empty disc space with bone chips (usually taken from the hip bone) or inserting two large screws through the facet joints. More latterly, metal cages have been used which though unable to absorb shock, at least preserve the disc space and prevent facet loading.

Disc removal has also been refined to make it less upsetting to the streamlined spinal mechanics. A microdiscectomy is a much more discreet surgical procedure and smacks less of removing the tyre and letting the wheel run along on the rim. This operation is carried out through a tiny cut in the skin and takes as little disc as possible (virtually the bulge only). With a smaller wound and less cutting, the scar formation is also kept to a minimum.

Apart from the essentials of skill, the most successful surgeons repair the surgical division made in the thoraco-lumbar fascia before they close up. This means the spine retains its ability to clench the vertebral segments vertically a few micro-seconds in advance of spinal action, which helps avoid instability problems developing later on. It also pays to keep the blood and oozing fluids to a minimum during the operation, and mop up as much as possible before closing the wound. Many surgeons demand the resumption of normal activity as soon as possible after a disc is removed. (The most successful surgeon I ever worked with, would not discharge his patients until they could touch their toes—usually ten days after surgery.) Early activity gets the working parts of the spine moving again, dispersing old bleeding and the collection of lymph in the tissues. In turn, this means that scar tissue is kept to a minimum with fewer 'adhesions' to clag the spine's delicate machinery.

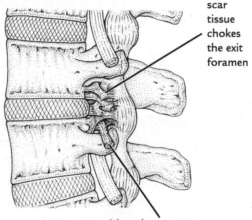

scar tissue chokes the exit foramen

withered nerve root

Figure 5.7 Scar tissue is like living undergrowth that invades the working machinery of a spinal segment and chokes the nerve.

The more selective surgeons have strict guidelines, and operate only if there is evidence of the nerves in the saddle area and legs not working properly. *Pain alone is no reason for opening a back and removing a disc.* Pain is subjective and too emotive to be any sort of guide. And besides, there are many other spinal disorders which can produce similar pain.

CAUSES OF A PROLAPSED DISC

- Pre-existing breakdown alters the properties of the nucleus and weakens the disc wall
- Bending and lifting stress breaks down the back wall of the disc

Pre-existing breakdown alters the properties of the nucleus and weakens the disc wall

Discs are shock absorbers and are meant to bulge. In their healthy state, the girth of each one swells imperceptibly as we transfer support from leg to leg during routine weight-bearing activity. As compression passes downwards through the spine, forces are transmitted outwards in all directions via the fluid of the disc's nucleus. Through the mechanics of an hydraulic sack, compression is converted to a springing-apart buoyancy which gives the spinal links their tensile strength and prevents the column juddering as we make contact with the ground.

As the spine sinks and springs with movement there is a synchronised to-and-fro exchange of energy. This passes between the briefly distorting nucleus and a moment later, the stretching mesh of the disc wall as it accommodates the force. When the wall nears its limits of stretch, its elastic recoil bounces the energy back to the nucleus causing it to puff up and thrust the spine aloft. This sublime dynamic uses the tension of the disc walls to off-load the weight of the spine bearing down from above and gives us a bouncing spring to our step.

The transfer of energy works well while the nucleus and the disc wall are in peak condition. As long as the nucleus retains its proper viscosity and the wall its tensile strength, a disc can dissipate on-off compression almost indefinitely. But early breakdown of facets or the disc—or overzealous muscle spasm protecting either—can change everything. Simple stiffness of the front or back compartment can eventually lead to disc prolapse, simply by localised muscle clench hampering disc nutrition.

As the disc dries out and the nucleus becomes more viscid, it is more easily deformed under pressure. Instead of being a tightly contained ball of fluid at the centre of the disc, the nucleus loses cohesion and spreads out laterally inside the disc when compressed. As it squirts this way and that with different activities of the spine, the nucleus bumps up against the internal layers of the disc wall. Over time, the battering against the interior walls amounts to trauma and eventually they start to perish from the inside out.

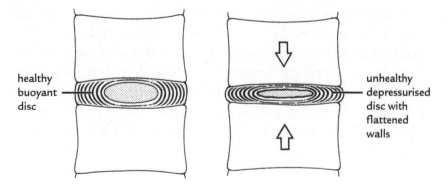

Figure 5.8 As a disc degenerates it flattens and the nucleus deforms laterally under pressure.

Muscle spasm at this point can convert a fleeting problem to a permanent one if the protective hold locks in and cannot be persuaded to disperse. The muscle clamp keeps the segment compressed in a vertical vice which hinders the dynamics of on-off squash and suck through the disc. As the compression persists, the disc wall starts to bulge at points on its circumference where pockets of escaping nucleus start pressurising the walls from within.

Activities that create sustained pressure inside the disc accelerate its breakdown. With generalised slumped sitting for example, when the pressure within the discs is raised, more force is directed towards the back wall of the disc. But with bending and lifting, which always involves some twist, the force is directed to the left—or right-hand back corner of the disc.

Bending and lifting stress breaks down the back wall of the disc

When the nucleus is on the run, ever-present bending of the spine has dire consequences. The intradiscal pressure is great when the spine bends forward. If there is a degree of twist as well, the pressure is even greater, because intense muscle effort increases the clamping-down pressure on the disc. If the twisting action is always in the same direction the nucleus damages the same part of the inner wall and breaks it up, strand by strand, like elastic fibres in a corset perishing at the points of greatest wear.

Strenuous lifting can be the last straw. It places enormous stresses upon the spine, particularly the lowermost discs. It raises the intra-discal pressures to unparalleled heights and as more fibres break in the same spot, a radial split opens up in the wall from the inside out. With more lifting and more pressure, the squirting nucleus burrows its way into the opening and prises it apart as it makes tracks towards the periphery. Eventually the whole wall may split open, disgorging its nucleus into the spinal canal. This is known as fenestration.

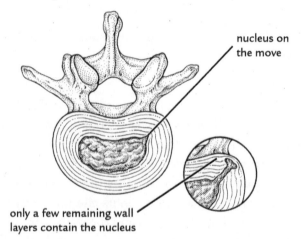

nucleus on the move

only a few remaining wall layers contain the nucleus

Figure 5.9 With torsional lifting strains, multiple fibre breakage at one point in the disc wall creates a small 'V'-shaped nick which the runaway nucleus burrows into.

Intensifying the breakdown

Rupture of the disc happens faster if the weight being lifted is held out from the body or if the lifting is heavy work—both of which increase the internal pressure of the disc. It also happens faster if you bend over using a large twisting element to lift. The disengaging of the facets as the segment goes forward makes the disc even more vulnerable to torsional strain and the alternating layers in the wall tend to separate, making circumferential tears in the outer zones. This appears to happen more readily in kidney-shaped discs, where the wall buckles around the sharper corners of the disc. With marked internal derange-ment of a disc, a radial split can meet up with a circumferential one and the nucleus can squeeze through many parts of the wall.

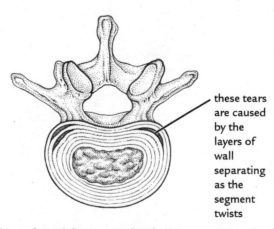

these tears are caused by the layers of wall separating as the segment twists

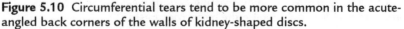

Figure 5.10 Circumferential tears tend to be more common in the acute-angled back corners of the walls of kidney-shaped discs.

The combination of the internal squirting pressure of the degraded nucleus and the rotatory strains on the wall means discs are most likely to disintegrate at the points on the clock face corresponding to 11 and 1. This explains the preponderance of postero-lateral disc bulges. In a good example of Murphy's Law, these two points of breakdown are exactly where the sciatic nerve roots make their way out of the spine, having branched off the spinal cord higher up. The nerves travel down inside the canal as multiple strands and then exit via the left and right intervertebral foramen at their designated level.

With a posterior protrusion the nerve is blighted inside the roomy central spinal canal, but a postero-lateral bulge irritates the nerve inside the much narrower exit foramen. The lack of room in the foramen means the nerve can get doubly assaulted; squashed up against the other wall while at the same time being stretched over the back of the bulge, like squeezing past the fat lady in the bus.

Not surprisingly, disc prolapse and fenestration is often brought on by strong physical work. Heavy lifting with a twisting action is the worst: moving bags of cement or digging with a long-handled shovel. Nurses are particularly prone to back trouble, though not always discogenic. Bad lifting techniques can 'hurt' the outer ligamentous wall but the disc will not prolapse until its nucleus has been rendered expressible by the pathological processes typical of a stiff spinal

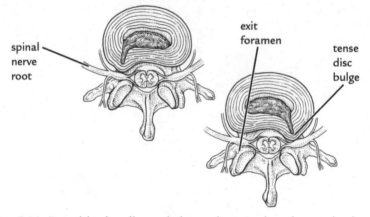

spinal
nerve
root

exit
foramen

tense
disc
bulge

Figure 5.11 Repetitive bending-twisting actions weaken the two back corners of the disc, right where the nerve root passes.

segment. Most commonly, nurses get a bad back when they save a patient from falling. This may create a virgin injury or, more probably, it strains a dormant link in the spine which was too stiff to accommodate the wrench passing through.

THE WAY THIS BACK BEHAVES

The acute phase

Sciatic pain usually creeps up over several days after hurting your back. You can often recall exactly what you did, though your back did not 'go'. You probably felt a slight strain, with a momentary deep pain inside which quickly passed off.

You often hurt the back by awkward lifting, where the object is unwieldy rather than unliftable. You may be hauling up one end of a sofa for example, and attempting to push it across the floor, when a corner snags. The difficult wrestling which follows is often the last straw, causing a sharp feeling of strain in your back. It becomes more sore and tight over the next few days, and then the pain down the leg starts.

The overwhelming feature is a painful tension deep in your buttock and down the leg which develops into a nasty cramp-like pain. In the beginning it feels like a pulled muscle or a tightly pulled string in the leg. It often originates in your buttock and moves down into the thigh, missing out on the knee and reappearing in the calf. You can usually locate a trigger point deep in your buttock by digging in through the layers of muscle with your fingertips, and for some reason, pressure here relieves the pain in the leg.

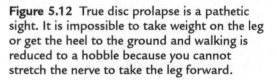

Figure 5.12 True disc prolapse is a pathetic sight. It is impossible to take weight on the leg or get the heel to the ground and walking is reduced to a hobble because you cannot stretch the nerve to take the leg forward.

The nerve may be so inflamed and sensitive to stretch you cannot even get your heel to the floor when standing. Your back often acquires a lateral 'windswept' bend (sciatic scoliosis) to relieve the tension of the nerve root and the whole back takes on a painful crippled appearance. Sometimes the sciatic scoliosis can be so pronounced, with the hips out one side and the shoulders the other, you look as if you won't fit through a doorway. From the rear view the spine looks completely hunched over and weak. As well as its lateral 'S' bends it will have a lumbar hump instead of a hollow lordosis. The buttock of the bad side can be flat and wasted, making a continuous line with the rounded low back. Both spinal deformities are protective mechanisms to minimise the tension on the inflamed nerve root.

If you have to stand, you rest your weight on your toe and keep the knee bent to avoid stretching the nerve, and the leg often shakes uncontrollably. Walking is reduced to a pathetic hobble. Every time you take a step forward there is a dreadful, mind-numbing pain down the leg like a jagged, red-hot spear through the tissues (in the textbooks this is described as 'lancinating' leg pain). Bending forward is just as impossible. As you attempt to go over, there is a crippling pain down your leg and the spine slews off more into its 'windswept' deformity to avoid the stretch.

Sitting is often unbearable because the bunching down of the spine increases the pressure on the disc and hence the tension on the nerve. After a few seconds the pain can be so bad you have to get up and lean on something to hang your leg to take the weight off. The pain can be just as unbearable after a few minutes of standing, when the pressure on the disc increases to a crescendo of cramping pain. (You hear of people on their way to being admitted to hospital having to lie down on the floor of the elevator to find relief.) Often, the most comfortable position is lying on your side in the foetal position with a pillow between the knees.

What causes the acute pain?

The back pain of acute disc prolapse is caused by the stretching of the disc walls and the pain genesis here is not dissimilar to that of stretching a stiff disc (see Chapter 2). The pressure of the localised bulge stimulates globular mechanical receptors between the fibres which manifests as a very uncomfortable pain deep inside your back, not relieved by the pressure of hands.

Only the outer layers of the disc wall have a nerve supply which explains why lesser bulges, which are so common, are invariably painless. These can be thought of as full-thickness bulges where the intact inner layers of the wall absorb most of the pressure of the nucleus discharging laterally. They take up the slack and save the more sensitive outer layers from the head of pressure caused by a runaway nucleus.

Once a degraded nucleus is on the move, it acts like a wedge which penetrates a small nick in the innermost walls and spreads it wider as it migrates to the periphery. As it gets closer, with fewer remaining layers left to contain it, the tension on the sensitive outer wall becomes immense, further fuelled by the compressive clamp of the muscle spasm. (This may explain the often-observed loud pop when a surgeon cuts through the wall of a sick disc during an operation, expelling nuclear material many metres across the room.)

As the condition worsens the old familiar back pain disappears as the leg pain comes on. This is usually caused by the disc wall breaking which relieves one set of problems but creates another. By this stage, the nucleus is a brownish colour (indicating it is degraded and toxic) and it is a potent chemical irritant on the nerve—especially if the nerve has already been sensitised by pre-existing stretch.

Stretching a nerve root is greatly more irritating than mere pressure. After all, we know from leaning on our funny bone at the elbow that pressure alone does not cause pain. It may temporarily lose conduction as the arm goes to sleep, and it is unpleasant and pins-and-needley while it wakes up again, but never raging pain. Pulling a nerve taut, thus subjecting it to friction as well as tension (as it twangs and rubs past other structures) is a much more potent irritant. Thus a lesser bulge which does not stretch the nerve will not be painful, explaining why such a high percentage of the population has disc bulges with no accompanying pain.

The first effect of pressure (and tension) on the nerve root is to restrict circulation. Fresh blood is prevented from squeezing through to the affected part, and stale blood damming up cannot flush the products of metabolism away. Both circumstances irritate free nerve endings in the local tissues which register another tier of discomfort from the problem area.

Remember, the inflammatory reaction does not apply to the disc itself because it is virtually bloodless. Rather, it applies to all the other red and swollen tissues clustering around the swollen disc and adding to the engorgement. This racks up the muscle spasm to another level, increasing the compression of the disc as everything becomes more inflamed and squashed together in a confined space.

When the nerve is under pressure and also stretched, friction develops between the tightly stretched nerve and its own protective sheath. The physical chafing between the two blood-congested (hyperaemic) surfaces causes much more pain as the inflammation of the nerve intensifies. Clear fluid weeps from the angry, raw surfaces such as you see from a burn of the skin, and the pain becomes unspeakably bad. If you could look inside you would see the nerve grotesquely red and swollen and the surrounding tissues swimming in fluid. This is the metabolic climate which causes agonising pain in the leg and which is very difficult to settle conservatively.

The disc, as the least bloody structure of the segment, is not a bad choice for surgical excision if the problem gets that bad. When everything is locked by irreversibly bloated congestion, this highly pressurised but inert component is the easiest to define and section out. It is a quick and effective way of decompressing a segment when all conservative methods have failed, despite the ill-effects this may have upon the future function of the spine.

The chronic phase

Rather than pressure, the chronic phase of disc prolapse is more about the internal machinery of the segment struggling with the after-effects of inflammation. As a legacy of the previous pitch of inflammation, the weeping exudate from the nerve gradually solidifies into thickened strands of scar tissue (see Figure 5.7 on page 141). This forms a matted mess, gluing the nerve to its sheath and other structures nearby, including the wall of the exit canal. Dry, whitish adhesions permeate everything, creating a choking collar which gradually strangles the nerve. This is called 'root sleeve fibrosis' and is very common after disc prolapse.

The tethering stops the nerve pulling and retracting freely through the bony tube (foramen) as the moving leg exerts traction on it. A dense undergrowth of adhesions dominates, often binding the nerve down to the back of the disc. The nerve is often much thinner than it should be because it has been trapped and choked for so long.

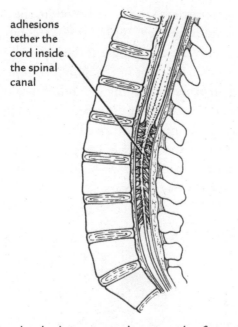

adhesions
tether the
cord inside
the spinal
canal

Figure 5.13 Dural tethering occurs when strands of scar tissue bind the spinal cord to the inside of the spinal canal, just like pegging Gulliver's hair to the ground.

Your leg can feel like a rigid extension of your back. It cannot bend freely at the hip to sit, nor make a step forward to walk, without taking your back along with it. This causes a typical waddling gait, where the low back rotates in a bottom-twisting fashion instead of your leg angling cleanly back and forth at the hip socket. Your back is stiff, with different pains according to what you do, and the pain down your leg comes and goes, depending on how much tension you put on the nerve.

Sometimes the spinal cord is tethered to the inside of the spinal canal. When sitting, the back is not free to slump and there is a dull pulling sensation which spreads high into your back and down into your buttock and thigh. This is called 'dural tethering'. The sitting stretches the spinal cord and tugs it where it adheres to the wall, causing a deep spreading pain which can extend right up to the shoulder blades. Sometimes you can almost feel the tightness inside the spine bowing you forward.

If the tethering is restricted to the nerve root in the exit canal, most of the symptoms are in the leg. With sitting, the buttock wants to come forward to lessen the acute angle at the hip, and the knee bends automatically when you attempt to straighten your leg. After a time, sitting may bring on other symptoms, like numbness of the heel, or a pain under the foot. But the tight dull pain in your thigh will always be worst because the slumped posture causes traction on the nerve root where it is glued to the foramen. Long after all other symptoms have gone away, a lengthy car trip or aeroplane flight can bring on a pain which you have not had for years.

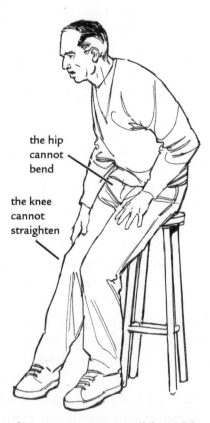

the hip
cannot
bend

the knee
cannot
straighten

Figure 5.14 A tethered nerve root cannot pull free of the exit canal as the stretching leg puts tension on the nerve. It is like a strand of cooked spaghetti adhering to the pot.

Apart from the cobbled-up leg and the difficulty sitting, there may be other subtle signs of the nerve not working properly. There can be a low-grade wasting of the muscles of the bad side. The buttock can look flat and withered, as can the calf, where the muscle has less tone and is floppy. There can also be less obvious signs, like the arch of the foot slowly dropping which causes the forefoot to splay out, sometimes making you feel your foot is too big for your shoe. You may notice that certain actions are weaker: you cannot stand on your toe or push off with that foot. When you walk, your leg feels heavier and more difficult to control and you have to haul it up when you take a step.

What causes the chronic pain?

Aggressive tension strains on the nerve, such as you might get kicking a football, can cause a localised inflammatory reaction right at the point where the nerve is bound down. Instead of the nerve pulling loose out of its exit canal, like a strand of cooked spaghetti, it is barely free to budge. The wrench may break a few adhesions and cause a small bleed in the otherwise whitish tissues, which then goes on to create more scarring and increases the cobbling effect. In the meantime there is more of the familiar pain in your leg, as the nerve is sensitised by the local inflammatory reaction.

Beyond a certain point, the invasive proliferation of adhesions may cause symptoms of 'vertebral stenosis'—internal narrowing of the spinal canal—because the nerve's own blood supply is hampered by the living junk crowding out the foramen. With this condition you feel pain in the legs whenever they have work to do. You often have to stop and sit down after walking a short distance, even sooner going up hills or steps.

Normally, when your leg muscles have to pump vigorously to move your body, the nerve sucks in blood to keep it relaying messages back and forth to your brain. If conditions are too cramped the nerve cannot puff up to get the blood through. As it suffers anoxia (lack of oxygen) your legs get more and more leaden, until a dreadful cramping pain locks them up completely and you have to stop. You have to rest by bending over or squatting, which broadens the inside diameter of the spinal canal and gives relief by letting more blood through. You can also get vertebral stenosis with facet arthropathy (see Chapter 3) where the arthritic bulk of the facet joint creates a similar effect on the nerve.

After a few moments, the pain eases and you feel more comfortable. When you resume walking, however, the pain starts up sooner and you have to rest more quickly. Each time you start, you travel shorter and shorter distances before your legs become painfully burdensome again and slow you to a halt. By the end of your journey, you feel you have barely started before you must stop. (This feature of shorter and shorter distances covered between respites distinguishes spinal stenosis pain from the cramping leg pains caused by circulatory problems.)

Although there is a tangible organic reason for the legs locking up in this way, it is fascinating how much this can change from day to day. Some days you can walk an entire block and the next you can barely get down the path to the street. The variable in the equation is the degree of muscle spasm in your back. When present to any degree it compresses the segment and further inhibits the blood getting through. In this regard, anxiety and tension also have a role to play because they influence the degree of tone in the muscles. If you are particularly tired or 'uptight', your legs will be harder to move and you will have the familiar wading-through-treacle feeling over the smallest distances. Other days, for what seems no good reason, you could almost be walking on air.

WHAT YOU CAN DO ABOUT IT

The aims of self-treatment of a prolapsed disc

In the acute stage of disc prolapse, the overriding concern is to gap open the back of the lumbar vertebrae to take the pressure off the bulge. This is achieved by rocking the knees to the chest, but benefits will be short-lived unless the muscle spasm can also be relaxed. This will not happen unless the inflammation of the soft tissues is dealt with. Medication prescribed by your doctor is necessary on both counts—NSAIDs (non-steroidal anti-inflammatory drugs) and muscle relaxants. Doing early reverse curl ups, even when there is severe sciatica, will also help ease the muscle spasm in the back.

As soon as the vascular engorgement is on the move and the inflammation of the nerve has started to settle, the important job is to seek permanent separation of the segments. (This may be thought of as the 'sub-acute phase' though I have not described this in detail in this text.) This is where both the BackBlock and the squatting exercises come in; to create traction which stimulates proteoglycans synthesis to enhance the disc's osmotic power, but also to physically suck in small quantities of water.

In the chronic phase the emphasis is both on segmental stabilisation and stretching. Sometimes segmental instability is waiting in the wings, brought on by lower hydrostatic pressure and weakening of the disc wall. Both toe touches and diagonal toe touches will suck fluid into the discs as well as strengthen the deep intrinsic muscles across the interspaces. The diagonal toe touches and the diagonal floor twists also stretch the adhesions in the exit canal which may be a legacy of past inflammation. The nerve root may be tethered to other structures nearby, and the rhythmic stretch and release of the nerve, pulling with the bending, helps persuade it free. At this stage, rotatory spinal movements also loosen the diagonal lattice of the disc wall, which frees it up to imbibe more fluid.

A typical self-treatment for acute prolapsed disc

Purpose:
Ease muscle spasm to relieve compression on the disc, pressure changes to stimulate disc repair, gap the back of the vertebrae to relieve the pressure on the disc, and disperse local inflammatory engorgement.

Rocking knees to the chest
(60 seconds)
Rest, lower legs on stack of
pillows for support

REPEAT 4 TIMES

Medication. Rest in bed with the lower legs supported on a stool or pile of pillows, hips and knees at 90 degrees. Repeat the knees rocking every 2 to 3 hours. See Chapter 7 for descriptions of all exercises, the correct way to do them and the reason for doing them.

For how long? Continue until the acute cramping leg pain has gone.

A typical self-treatment for sub-acute prolapsed disc

Purpose:
Ease muscle spasm to relieve compression on the disc, pressure changes to stimulate disc repair, gap the back of the vertebrae to relieve the pressure on the disc, disperse local inflammatory engorgement, break up spinal impaction from muscle spasm, and strengthen tummy to lift weight off disc.

Rocking knees to the chest
(60 seconds)

Rolling along the spine
(60 seconds)

Reverse curl ups
(five excursions)

REPEAT ALL THREE EXERCISES 3 TIMES

Treatment sessions should be twice daily, morning and afternoon, followed by rest, with the lower legs supported, for 20 minutes. When you are up and about, avoid lengthy periods of sitting and standing still, and try to have a short walk each day (less than 15 minutes), walking briskly and lightly with your tummy and pelvic floor drawn up. It is ideal to progress to having two short walks per day, spending most of the day lying down.

For how long? Until severe intermittent leg pain has gone.

A typical self-treatment for chronic prolapsed disc

Purpose:

Ease muscle spasm to relieve compression on the disc, pressure changes to stimulate disc repair, gap the back of the vertebrae to relieve the pressure on the disc, disperse local inflammatory engorgement, break up spinal impaction from muscle spasm, strengthen tummy to lift weight off disc, decompress the spine to aid disc regeneration and repair, stretch adhesions, and restore coordination between the tummy and the two groups of back muscles.

Rocking knees to the chest
(30 seconds)

Squatting
(30 seconds)

BackBlock routine   Step 1   (60 secs)

Step 2   (30 secs)   Step 3   (15 times)

REPEAT THESE THREE EXERCISES 3 TIMES

When you no longer have *leg pain* you can include the following two exercises in your routine:

Floor twists
(four times to bad side;
one to good

Diagonal toe touches
(down to each foot 3 times)

Repeat three times per week *only*. You may have vague leg pain after these sessions but it should disperse before the next one. If it does not, you must revert to the sub-acute stage regimen.

For how long? Continue indefinitely.

6

The unstable spinal segment

This is the end condition in the breakdown of a segment when it becomes like a weak link in the spinal chain.

WHAT IS SEGMENTAL INSTABILITY?

An unstable spinal segment is the antithesis of a stiff spinal segment. It comes about when the two main stabilising structures of a motion segment—the intervertebral disc and facet joints—become stretched and weakened, usually as a result of the degenerative process. Full-blown segmental instability is rare, and can be testing to cure by conservative means because so many systems are in trouble at once—all fuelling one another—and the sources of pain may be multiple.

There is always fleeting 'muscular' instability brought on by temporary under-activity of the deep spinal muscles whenever a segment is inflamed. It is a type of reflex inhibition which switches off the local (intrinsic) muscles to spare the painful segment from additional compression, but it causes mischief when it exposes a structurally weakened segment to excessive movement.

This is the usual route a stiff segment takes when it becomes unstable, and though it may never eventuate, all stiff segments are vulnerable to this course of events unfolding. Once symptoms of frank instability have developed, they can be kept under control as long as the deep-acting segmental muscles do their job in keeping the weak link secure. From this, you can see how imperative it is to keep

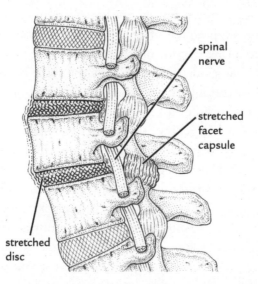

spinal
nerve

stretched
facet
capsule

stretched
disc

Figure 6.1 When the fibrous union of both disc and facet capsules becomes stretched, the segment must rely on the primitive bony notching of the facet joints and muscle power to keep itself in place.

spines working properly once they have developed bad movement patterns from pain. You can also appreciate the scope of intrinsic exercises, through all stages of spinal dysfunction, in keeping instability problems at bay.

Segmental instability develops when primary weakness of the disc eventually translates across to the facets, or when laxity of the facets translates back to the disc. Both structures share almost evenly the job of gluing the spinal segments together and holding them secure when the spine bends. If one fails, greater responsibility is transferred to the other. As soon as the segment is loose in both front *and* back compartments, it can jostle around in the column with only the primitive bony notching mechanism of the facets and the power of the intrinsic muscles to keep it in place.

The confusing thing about instability is that it is often hard to pick up. The defensive response from the spine can be so total that your back can simply feel rigid. It is usually impossible to sense a loose segment because they all jam together en masse and the spine operates as a splinted rod. The weak link may only come to light when the spasm starts to fade and the spine goes to bend. As it goes over without being properly braced, the loose vertebra goes to slip forward, like a drawer slipping out of a chest of drawers as it tips over.

There are other circumstances where a loose vertebra moves, in a much more low-grade way, every time the spine bends. This may not be especially worrying; you may sense a tiny click or slip in your lower back as you go over, or a small wriggle in the movement, which you cannot prevent happening. In other circumstances, there may be a small arc of pain, just after your spine leaves the vertical, whereafter it moves freely and you can go right to your toes. Returning to upright, there is a similar painful phase in the movement, just as you near the top. This is exactly the point where the segment slips minutely.

Sometimes your spine continues for months with the segment slipping this way every time you bend. There is always a degree of background grumbling, achey stiffness but if you do something to hurt your back, it will suddenly stiffen more. As the muscles become more rigid, your back takes to moving more awkwardly and the clicking gets louder—until the stiffness becomes so limiting there is very little movement at all, and the clicking then stops. At this point, depending on other circumstances (such as how tired you are; whether you are unwell) the defenses of the spine can reveal themselves as not up to the job of protecting the weak internal link. If an action is awkward as well as weighty, or if there is another mishap as you go to bend, you will have an ominous sense of your spine starting to give way. Before you can stop it, you are caught and your whole back collapses, doubled over like a broken reed.

Although the degree of uncontrolled movement of the vertebra is minuscule, it still constitutes 'unstable' activity. The micro-trauma from the repetitive slippage and the giving-way incidents all add up to inflame the structures trying to hold everything in place.

The aberrant movement is usually forward glide (segmental shear), because all lumbar vertebrae have scant restraint for controlling this. Usually, the forward tipping of the vertebra takes up the slack in the posterior ligamentous system, and in this way forward shear is controlled. As the vertebrae tip, their aft section lifts away from the neighbour below which disengages the bony facet lock. This frees it up to glide forward more until the two facet surfaces butt up against one another once again and the lock re-engages. In the clinical setting we simply control shear by building up the muscles that control tip, and by this I mean multifidus.

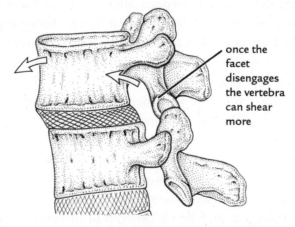

once the facet disengages the vertebra can shear more

Figure 6.2 We have few soft-tissue restraints for controlling forward shear of the lumbar segments—only the bony block of the facets engaging. To invoke a soft-tissue brake earlier in range we can control shear by humping the back to control tip.

When the loose vertebra goes to slip, there can be a feeling of the weak link about to give way. This usually happens when the spine moves forward from the vertical position, like a stack of children's building blocks coming undone, as it leaves the relative stability of the vertical. It can also happen when the spine is stretched across in the slung-out position, such as when leaning across to make a bed.

It is usually a lapse in the guarding role of the tummy and spinal muscles, and then their too-late over-reaction, that brings you undone. They all clench instantaneously which takes your breath away, but unlike the out-of-the-blue facet locking episode described in Chapter 4, you have a sense of the familiar; you feel danger coming and you can stop it before it reaches the critical point. The muscles growlingly go on guard, flickering in a menacing sort of way as you veer into your 'weak territory'; a warning sign to stop the movement and backtrack out of it before you go too far.

If time passes without the back folding up under pressure, the incremental slippage can stop happening. One day you realise the stiffness has gone, and the clicking beneath the surface has also faded.

Your back moves better as the harmony of the muscles returns. This usually happens with a fitness or weight-control binge, particularly if it also involves tummy strengthening. It means the weak link has made itself more secure by becoming stiffer, or the segmental muscles across the link have taken to working better. This precarious truce is more or less maintainable (unless you do heavy pushing work), but the former rumblings cannot be wholly ignored. The weak link will always be the first to give way when your back is next put to the test.

When the weak link is stressed, pain may not come on immediately; returning stiffness may be the first clue. This gets more and more noticeable a few days after chopping down a tree, or taking a car trip over a rough road, and then a gnawing pain starts in the leg. The weak link is more susceptible to knocks and bangs passing through the spine, and slowly the level of inflammation rises as the segment is squeezed and the disc is 'milked' by the tightening muscles.

Diagnosis

Diagnosing full-blown instability can be difficult because of the complexity of the pain picture. There can be so much pain, it is hard to know where to start. Action X-rays rarely show the segment opening and closing more than it should do because the spinal defenses are much too clever for that. The surrounding muscles splint the loose vertebra and make it appear stiff. A discogram can show internal degradation of the disc, but the most definitive sign of instability is none of these: it is the presence of tooth-like extensions of bone (osteophytes) around both the interbody and facet joints.

With longstanding instability, the body throws out extra bone around the disc perimeter where it meets the vertebral body. The bony spurs provide a broader base of bone for the disc wall to take hold, creating additional surface area for the individual fibres of the disc wall to tie themselves to, thus providing stronger anchorage for the stretching wall.

Similar changes occur in the facet joints when the primary instability is there. The lower facet surface remoulds itself to make a more enveloping lip of bone which curls up around the upper facet surface to hold it in place. In the business, these are called 'wrap around bumpers' and they too are thought to be an ingenious attempt by the body to make the joint more secure. The bony cupping traps the upper vertebra and thus reduces its ability to slide around in-joint.

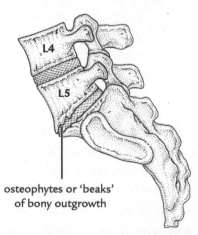

osteophytes or 'beaks'
of bony outgrowth

Figure 6.3 Claw-like bony outgrowths (osteophytes) of the vertebral body indicate instability of the front compartment whereas wrap-around bumpers of the facets indicate bony re-moulding in response to instability of the back compartment.

With instability problems, a great deal of focus must be given to re-educating proper free-flowing spinal movement, in case some freak action creeps in under the spine's guard and wrenches an unprotected weak link. This will create an unstable segment where none existed before.

In fact, that is the central aim of this book: preventing a stiff spinal link becoming unstable, or better still, preventing a stiff link developing in the first place.

Not uncommonly, the segment soldiers on valiantly with its chronically stretched ligaments, coping quite well with the anomalous movement of both front and back compartments. Often, it is when you do something extra to hurt the back that things flare up; the facet swells more or the bulging disc starts impinging on the nearby nerve root. Savage, intolerable leg pain often brings things to a head—and often to the point of a surgeon's knife.

Spinal surgery

Operating on spinal instability involves surgically joining the loose upper vertebra to the lower one by inserting two large titanium screws through both facet joints and then packing out the evacuated disc space with bone chips taken from the pelvic bone. This is called a spinal fusion. It is usually done after first removing the flaccid disc (and sometimes part of the facet joint) in order to relieve the pressure on the spinal nerve root. These procedures are called discectomy and partial facetectomy.

More recently, a whole slew of less invasive operative techniques have come into vogue to provide sorts of quasi-fusions. They involve using plastic or metal struts to join together either the transverse or spinous processes of two segments to limit excessive movement (and compression) rather than obliterate movement altogether, as the older fusions did. Though their rationale may be straightforward in controlling segmental participation, I have doubts about their efficacy, mainly because their alignment in situ has difficulty in controlling forward shear.

Like all operative procedures, I feel much more stringent selection criteria should be in place with proper biomechanical analysis and diagnosis before patients are shepherded this way as a treatment option. Unfortunately, though not the case with all surgeons, these devices seem to be the flavour of the month at the moment, whatever the spinal pathology and we need to see sound evidence for their deployment.

Furthermore, limiting segmental movement in this way brings almost to a halt any possibility of regenerating disc health by conservative means (because disc nutrition requires as much 'good' segmental movement as possible) so patients have to be sure that the slower route of regenerating disc health has been fully tried and tested before being discarded for the quick fix. And indeed, if surgery is to be contemplated, patients need actual figures of a surgeon's experience in using the device or procedure and the outcomes, without feeling difficult or demanding.

As you might imagine, surgical technique is of the essence with any spinal operation (I liken it to using a hammer and chisel on a

Stradivarius violin) because the spine is never quite the same after-wards; it is hard to 'go back' and conservative treatment is never quite as effective.

Apart from disturbing the delicately poised spinal mechanics, prolific scar formation causes many problems. If the scarring becomes invasive, it can be just as space-occupying and obtrusive as the structure deemed worthy of removal. In particular, the nerve root can be slowly strangled by the growth of adhesions, eventually causing the same symptoms of pressure on the nerve, and the old pain starts up again. Post-operative adhesions are similar to the post-inflammatory condition called 'root sleeve fibrosis' described with facet arthropathy and chronic disc prolapse (see Chapters 3 and 5).

The other complication of spinal fusion is the strain translated to the next working level up (L4 if L5 has been fused, or L3 if L4 and L5 have been fused). Both are almost flimsy compared to the robust L5–S1 junction and are ill-equipped to act as the seat of spinal move-ment. They are not bedded deeply in the pelvis like L5, nor do they have the august ilio-lumbar ligament to lash them down. Thus they are progressively over-taxed by routine movement. The problem usually takes several years to manifest and affects L3 more seriously than L4. Intrinsic spinal strengthening is therefore a critical part of a post-fusion regimen.

CAUSES OF SEGMENTAL INSTABILITY

- Primary breakdown of the disc
- Primary breakdown of the facet joints
- Incompetence of the 'bony catch' mechanism of the facet joints
- Weakness and poor coordination of the trunk muscles

Primary breakdown of the disc

In the later stages of breakdown of a stiff spinal segment, the disc degrades to such a degree that it becomes like an inert flattened washer. It dehydrates and loses its buoyant ball of nucleus at the centre which should act as a pivot. The nucleus so lacks internal hydrostatic pressure that it cannot prime the disc and spring-load its vertebra on top to tip as the spine bends. Instead, the vertebra shears forward. The passive non-performance of the disc tugs and stretches the side walls, similar to the way a perishing car tyre veers over onto its wall as it rounds a corner. Thus a flaccid degenerating disc is gradually destroyed by its own vertebra's runaway movement.

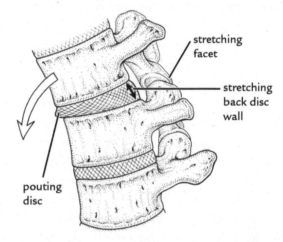

stretching
facet

stretching
back disc
wall

pouting
disc

Figure 6.4 Lack of internal pressure in the sick disc means the spine skids and slews forward on that level whenever you bend, instead of tipping on a buoyant nucleus.

When the disc has reduced stuffing, the constant weight of the spine bearing down through the weak link is further cause for break-down. With its much reduced intradiscal pressure, the main brunt of load-bearing is transferred from the nucleus to the walls. They have nowhere to go but billow outwards, like a cardboard box crumpling when a weight is rested on top.

With long periods of sitting, the distension of the walls becomes more marked because a degenerated disc loses its fluid more quickly. As the upper vertebra settles closer to the lower one, the disc bulges circumferentially like an inflatable rubber collar, and the two vertebrae almost touch. The back may be caused to slew sideways on the sick disc by the muscles attempting to lift the pouting wall off the nearby spinal nerve; it may remain kinked into a 'wind-swept' 'S' bend after rising from sitting, and take a few minutes to disappear. A sciatic scoliosis like this is often more long-lasting after attending a seminar for a few days or after a busy period of sitting writing reports.

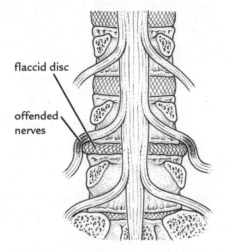

flaccid disc

offended nerves

Figure 6.5 With the disc bulging all around like an inflatable rubber collar both nerve roots are vulnerable.

This process of breakdown can be greatly speeded up by a percussive injury through the length of the spine as if the degenerated nucleus is looking for a way out. The sudden impact can blow a vent through the thin cartilage of the vertebral endplate causing the degraded nuclear material to squirt into the blood-saturated honeycomb bone (spongiosa) of the vertebrae above or below. Endplate fracture when the disc is healthy will not result in the nucleus breaking out, although it will cause rapid deterioration of the disc's health so that eventually it will punch through.

Once the nuclear material has penetrated the bone, the blood cells in the spongiosa attack the toxic disc material as an unwelcome foreign substance, and an auto-immune reaction sets up. This does not limit itself to the dislodged material but continues back through the communicating hole into what is left of the disc. The remnant disc is then devoured by the auto-immune reaction, leaving a flaccid fibrous bag of scar tissue where the disc once was. This process is called *primary disc disease.*

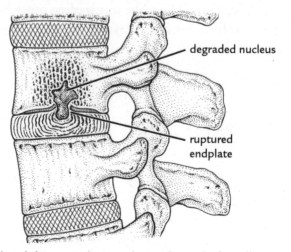

Figure 6.6 Disc breakdown can be greatly accelerated when direct vertical impact through the spine extrudes a semi-degraded nucleus into the neighbouring vertebral body.

Not surprisingly, this is a back problem common to the armed forces. Soldiers who train over assault courses are particularly susceptible through landing on their feet with a heavy pack on their back. Parachutists are also at risk, as are pilots who use ejection seats: they all suffer similar traumatic shocks up through the spine which may be painless at the time. It is also the complaint of the weekend gardener. Forceful tugging at a stubborn root, or unaccustomed heavy lifting, can punch a hole though the cartilaginous endplate which is insufficiently 'seasoned' to tolerate the force.

As the disc becomes incompetent, the segment will only remain stable in the column as long as the other main control mechanism of the facets is up to the task. The strain will eventually be felt on the facets and their highly innervated facet capsules are the first to complain.

Primary breakdown of the facet joints

Segmental instability can also start off when facet joints develop severe arthritic change (see Chapter 3). With poor joint lubrication and the friction of the joint surfaces chafing, the cartilage buffer within the joint can be worn down, leaving room for the two opposing bones to clatter about more as they lie alongside one another. This also causes the joint capsule to pucker and the joint develops excessive play during activity—even though in time it ingeniously moulds its lower bone surface into a cup-shape to hold things stable. Inexorably, the joint slips around with movement and articular destruction picks up apace.

Sometimes, instability can spread from repeated facet locking episodes (see Chapter 4). With each mini-dislocation, the capsule is traumatised and healthy fibres replaced by scar tissue. As the capsule becomes weaker it also loses elastic recoil, thus making it harder to keep its joint snugly together. However, facet locking typically affects one facet only, so instability is less likely to spread across the whole segment from this problem.

Incompetence of the 'bony catch' mechanism of the facet joints

You can also develop segmental instability when the solid bone-to-bone backup of the facet joints becomes incompetent. This can happen in three ways: a congenital malformation (called spina bifida) where the facet joints fail to develop properly in utero; a physical break in the bony neck at the bottom of the catch mechanism, usually caused by trauma (called spondylolisthesis); and when the bony neck of the catch mechanism very gradually elongates, like toffee stretching (called spondylolysis).

In all cases, the 'bony catch' of the facet's locking mechanism fails to prevent the upper vertebra slipping forward on the lower one. Without the lock of bone against bone—like a row of fingers of one hand hooked up against the same of the other—both the disc and facet capsules slowly stretch, letting the upper vertebra gradually slip over the abyss.

As dramatic as a spondylolisthesis can look on X-ray, with an overhang of sometimes as much as half the upper vertebra, it can be quite stable. The flattening of the disc as the vertebra pulls forward causes the disc wall to harden as it bunches down. Barring some fluke additional mishap knocking it loose (like a hard fall on the bottom or a scrum collapsing in a rugby game) the segment may remain symptom-free for years.

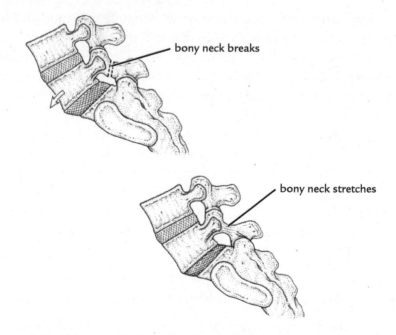

bony neck breaks

bony neck stretches

Figure 6.7 With spondylolisthesis, a break in the neck of bone below the lower facet surface can render the bony lock incompetent. With spondylolysis, the bone stretches over time rather than breaks. Both conditions can make a segment unstable.

I see this in clinical practice all the time: people with a low-grade backache unwittingly harbouring a fairly advanced slip, to have it flare up only when they suffer some minor new mishap. The recent trauma may cause an additional slip of a millimetre or two but even so, the upper vertebra usually settles down securely on the lower one quite quickly. The lesser symptoms of a 'stiff spinal segment' then return.

It is not unusual to break the neck of a facet with impact through an over-arched lower back, and we see this often enough with fast bowlers landing heavily on their leading leg at cricket. You can also break the catch mechanism with a fall on your bottom, and an astonishing number of the Inuit population has spondylolisthesis from falls on ice. Some of these cases remain chronically unstable, with fractures on both sides of the bony ring failing to unite. The back then permanently harbours a low-grade ache and a tendency to give way when caught off-guard or with an awkward movement. Even so, broken bones and all, it is often surprising how little pain there is.

Weakness and poor coordination of the trunk muscles

When an intervertebral disc dries and flattens due to the degenerative process, the intrinsic muscles which work the upper vertebra have a harder job. The angle of pull of their muscle fibres is flattened, which means their already poor mechanical advantage is reduced further. Thus it is harder for muscular control of the segment to compensate for weakness of the other two stabilisers, and the vertebra is more vulnerable to slip.

Full-blown instability can also develop from fleeting 'weakness' of a segment because the back is suffering pain; a temporary alteration in muscle balance can cause the segment to loosen more quickly. This happens for two reasons: volitional (conscious) and automatic. The *volitional element* is because of the way you use your back when it hurts—avoiding bending and making your knees do all the work.

People usually try to bend with their back straight and knees bent because they think this is the right way. Over the years this has been reinforced by old-fashioned back-care programs, in the belief that it puts less pressure on the discs. We now know this to be wrong and, furthermore, that using the back this way only makes matters worse. It emphasises the use of the stronger, clumsier muscles of the spine while under-utilising the smaller, fine-acting ones which control the individual segments. Thus the spine operates en masse in a bulky, non-bending fashion, and loses subtle undulating motion throughout its length. This makes it feel precarious in complex postures and safe only when it does simple, almost robotic actions. As it gets stiffer and weaker, it becomes increasingly vulnerable to shearing strains.

Even preliminary efforts to get the spine bending properly at this stage improves things—and despite your wariness, you often find it happens quite easily. After bending two or three times (preferably using your hands to walk down your thighs with the knees bent like a gorilla) your back immediately feels freer and more supple. You might easily get a sense of how bending might help you!

The *automatic element* of muscle dysfunction is a little more complicated and is more likely to happen if there is inflammation in the back. If there is an irritable disc or facet joint, the finer acting deep muscles can be 'paralysed' by the dominant activity of the over-protective long back muscles, even though their main purpose is guarding. But the finer group can also inhibit as a pre-emptive measure, to spare the inflamed segment excessive compression of the muscle working normally.

Some speculation

It is possible, though not documented, that transversus abdominus, the deepest acting tummy muscle, reduces activity if there is primary inflammation of a stiff spinal segment—because its action clenches the 'cotton reels' together vertically. Lighter neurocentral compression may lessen the pain, even though it ultimately hastens instability.

On the other hand, multifidus may automatically dampen its activity if there is a primary problem of the facet joints. Although this spares the irritable facet from compression, it too inevitably results in poorer control of the vertebra's tail and makes it more vulnerable to shear and rotational instability.

Whether, first up, there is over-activity of the dominant group or automatic inhibition of the deeper ones, there is no doubt that one fuels the other. The stronger group gets stronger and the weaker one gets weaker.

This is the way it works: whenever there is pain in the back, the long erector spinae back muscles develop 'protective reflex spasm' which splints everything semi-rigid and limits movement. They lock up the length of the spine, like guy-ropes holding a flagpole steady in a gale, jamming its base into the ground. As the low back compresses more, the intrinsic muscles relinquish their hold. This leaves all the spinal segments—but particularly a problem one at the base—more likely to shear when the spine bends.

The mechanism also works in reverse. Once the unstable segment reveals a tendency to shear, a self-fuelling cycle sets in. As if sensing the weak link, the erector spinae clench the spine even more. This makes the whole back board-like, but further disables the control of the segments and a cycle gears up. The spine develops a brittle veneer of stiffness which makes it awkward to move, and the weak link inside ever more prone to shear.

Patients often half recognise this themselves, with their back becoming increasingly sore and stiffness extending right up (people often complain that their neck is starting to click). It also keeps giving way at the base if they don't protect it. They lose confidence with anything involving bending, fearing they might end up in a heap on the floor, and they become hidebound by a multitude of constraints. They creep about, taking the path of least resistance and planning every move, even down to picking up a teacup.

THE WAY THIS BACK BEHAVES

The acute phase

Acute segmental instability is arrived at after being chronic for some years and suddenly taking a turn for the worse. Being unfit, putting on weight and sleeping in a very soft bed are background factors which together can tip the balance, but strong physical exertion can often be the final straw. Strenuous pushing activities—such as a car, wardrobe or lawnmower—are often linked to the back getting worse, no doubt from the backward shearing action of the spine on the flaccid disc which it is ill-equipped to prevent.

Then the variety of symptoms—and the emergence of new ones—can make the problem very difficult to unravel. Except for extra rigidity in the back, its worst features are often indistinguishable from acute disc prolapse, with an angry pain and paraesthesia (disturbed sensation) down the leg. What differentiates instability is the long history leading up to the present crisis.

At the peak of the acute phase, the erector spinae stand up like cables, making it impossible to hump your low back to bend. Try as you might, the muscles will not let go. Everything appears stuck, like a puppet with its strings pulled too tight, and then as you get further forward you elicit a crippling pain down the leg as the tension stretches the sciatic nerve. At this point, the back will be doubly disinclined to bend: both to ward off any slippage at the weak link and to avoid further stretching the inflamed nerve root.

You will be uncomfortable in most positions, even turning over in bed, because your spine has lost its serpentine control of all the segments as it undulates across the mattress. You will know from experience the best way to do this is either rolling over like a log, or very slowly with the tummy braced, so the spine does not 'disjoint' at the weak link. But if the instability is severe and the leg pain intractable, you will hardly know what to do with yourself. You will be pathetically crippled by your leg, and only comfortable lying in the foetal position with a pillow between your knees. Sitting may be impossible and walking may be reduced to a shuffle by the nerve root's painful inability to take stretch (see Chapter 5). Invariably at this point the condition needs surgical intervention.

What causes the acute pain?

With acute instability, there will be intense back pain, quite probably from more than one structure. It is arguable whether more pain will come from the side of the motion segment which became unstable first, although in the early stages the facets—with their propensity for inflammatory response and their sophisticated nerve supply—will always register a lot more pain than a disc. Broadly speaking, the true picture of instability is often a time-lapsed composite of the different conditions which have afflicted the segment during the course of its breakdown.

At one and the same time there may be symptoms from the stretched disc wall (see Chapter 2), from the inflamed facet joint (see Chapter 3), or from a localised bulge in the wall (see Chapter 5). As a result, there may be deep central back pain in addition to a pain spreading over the buttock. There may also be sciatica, numbness and weakness in the leg from both the capsular inflammation and disc prolapse. And just to complicate matters, there may also be referred pain in the leg from the facet joint arthropathy. It certainly can be complicated.

The sub-acute phase

If there is no sciatica it is easier to recover from this problem without surgery. If you persevere, your back will grudgingly regain segmental strength and learn to bend, especially if you pull your tummy in and hump the low back first. Soon enough, you will notice a tell-tale wriggle or kinking movement in the lower back emerging through the stiffness, even when your back is still quite cast.

The wriggle may come about as your spine tries to avoid skidding on the flaccid disc. Watching the movement happening from behind, the spine appears to do a lateral semicircular movement as if attempting to roll around the rim of the flattened disc as it goes forward. It can go around either the left or right side of the rim but either way, sometimes the aberrant movement can be more unnerving than painful. Until you build up the strength of your intrinsics, you can minimise this by pulling your tummy in hard and drawing up your pelvic floor, both before you bend over and as you come up again.

With the widespread armour-plating of the muscles you often feel more pain in your upper back as stiffness spreads to the better-off regions. Sometimes there is soreness right up to your shoulder blades and you may also complain of headaches. Sufferers often seek spinal manipulation at this point, which is ill-advised. Although manipulation comes into its own with other spinal conditions—particularly segmental stiffness if the vertebra is twisted on its axis, or with acute facet locking—there is no way, for even the most experienced operators, to accurately localise the manipulative thrust to the stiff segments while sparing the weak one nearby.

The so-called 'million dollar roll' (where the patient is put on his or her side and the low back clicked one way then the other) often wrenches the weak link, even though the sense of release which coincides with the characteristic popping sound can be profound. People usually get increasingly uneasy about having the same treatment over and over again when there is no long-term improvement. With manipulations becoming more frequent, and the benefit shorter-lived, many decide to look deeper and search out a more permanent (if not instantaneous) solution—and indeed to have some understanding of what is wrong in the first place.

You are usually at your most comfortable sitting, because the reduced hollowing in the low back lessens the forward inclination of the lumbar vertebrae and thus the tendency for the weak one to slip forward. But rising from sitting to standing can be agony. There is often a painful catch as you straighten and you may have to push yourself up through the final part with your hands on your thighs. Basically you will be unhappy about feeling the movement of the vertebra, and there is good reason for this. You will always be on guard against it letting you down.

You will know from experience that the more clicking and grating there is in your back, the stiffer it will become. As the back tightens, there is a dawning of the familiar pain down your leg, which starts off as an ache and then turns into a cramp. Sometimes the leg pain arrives before your back stiffens but either way, the re-emergence of familiar symptoms is a sign the link is inflaming again. It usually corresponds with a period of excess, when your exercise routine has lapsed or you have put on some extra pounds. Bear in mind too that the pain can also intensify if you do too much spinal strengthening (see 'Spinal intrinsics strengthening' Chapter 7).

As the back re-stiffens, you find it increasingly awkward to do everyday things. You may notice your actions becoming more laboured before the pain becomes obvious. Even something as simple as picking a belt up from the chair can become a farce. Your back stays ramrod straight and you feel safer bending sideways with your bottom out and bending your knees, rather than going forward normally.

The chronic phase

You can cope with a chronically unstable back indefinitely, if you can stop the weak link inflaming. You are usually the best judge of the way to do this, as you are more in tune with your back than most practitioners. Although pain may not be a big factor with the low-grade instability, you are strikingly susceptible to minor mis-adventure. Strengthening of the weak link is imperative, to stop you being catapulted into difficulty at a moment's notice.

With chronic instability, there is muscle stiffness all over the back and a soreness which is pretty well localised to the problem level. However, the level of pain is unpredictable and varies with what the back has to endure. Something as simple as stepping off a curb can set it off—because the lack of support allows the spine to strain at the weak link—but so can heavy jarring exertions which rattle the link.

When the condition is truly quiescent, the back rarely gives way when you go to bend forward and you hardly ever feel it clicking and grinding. You can set off a peaceful back however (and you will be annoyed with yourself that you should have known better) by over-doing sustained bending activity, such as vacuum cleaning or laborious sweeping or raking, or pushing anything heavy, like a wardrobe. (It is always better to turn around and push with your bottom.)

Difficulty with straightening after bending is a sure sign things are wrong. Because the pooling of interstitial fluid causes a transient wedge of waterlogging at the back of the over-mobile segment, it is difficult to close the gapping at the back of the spine to let you straighten. For the first few steps you get about like an old person with knees bent and bottom out, and both hands on the back of your hips to help winch you straight.

More often than not it amounts to nothing more than that. Your back feels normal again within a few steps and the stiffness fades. But occasionally you stay doubled over and a generalised stiffness sets in across the back which you cannot throw off. It can linger for several days, making it increasingly awkward to do things—until you do something thoughtlessly, your back gives way, and you are tipped into an acute phase again.

Although the mishap usually happens when your back is caught off-guard, it is unlike acute facet locking (see Chapter 4) in that you can sense your back getting ready to slip at the weak link and you can avoid it by stopping the movement short. Sometimes if you are quick enough you can stop yourself collapsing to the floor by getting your hands to your thighs before your back gives way. Then you can unfurl yourself to vertical by thrusting your pelvis forward under the spine and pushing up with your hands.

You will see from the self-treatment section how important it is to stop the spine stiffening up all over at this point; it is essential to prevent your back getting so rigid that the deep muscles automat-ically inhibit and you lose control of the individual segments. As soon as possible you have to get down on the floor and roll back and forth along the spine to prevent the segments jamming up.

What causes the chronic pain?

The pain of chronic instability is probably caused by a combination of the micro-trauma stretching the fibres of both disc and facets, and the compression of the runaway segment by the spine's defense mechanisms. In practice, however, it is impossible to differentiate between the two.

As fibres right across the motion segment are progressively damaged by the excessive movement, toxins are released which stimulate free nerve endings, thus alerting the brain that all is not well. This manifests in the form of pain—both from the disc wall and the facets—plus a greater or lesser degree of muscle spasm. The spasm is not only painful in itself but contributes in its own way to more instability.

Meanwhile, there can also be pain from the vascular engorgement of the weak link when it is locked up tight by the large muscles of the spine holding the whole back stiff. When the cable-like muscles clench all the segments together, as if they were in a mechanical vice, the weak problem link in the middle can bloat up painfully because the circulation through it is hampered by the compression.

Then, you not only get pain from the oxygen deprivation (anoxia) of the tissues (because there is no fresh blood coming through to bring more oxygen), but you also get pain from the rising concentration of metabolites or waste products in the tissues because the old blood cannot get away.

WHAT YOU CAN DO ABOUT IT

The aims of self-treatment for segmental instability

The immediate aim when treating acute instability is to dampen the inflammation of the weak link so the nerve settles down. Initially, this involves relaxing the muscle spasm to let the pressure off the segment and allow a better circulation to pass through. Treatment in this phase is exactly the same as for the acute disc prolapse, with an emphasis on rocking the knees to the chest to open the back of the spine, which also starts introducing pressure changes through the flattened disc. It is then important to establish better movement patterns as soon as possible, so the spine does not keep skidding on the flaccid disc. Reverse curl ups start to strengthen the tummy and help switch off the long back muscles.

As soon as the pain is on the move, it is important to start treating the back 'normally' so the intrinsic muscles play a more active role knitting the loose link together. The natural ebb and flow of movement also helps pump the congestion away and stimulates proteoglycans manufacture within the sick disc. Toe-touching exercises should start as soon as possible so the spine gets accustomed to bending with segmental control, and together with squatting and the BackBlock they help puff up the disc to make it more tense. Encouraging the flattened disc to imbibe fluid not only takes pressure off the walls but gives the intrinsic muscles a better angle of pull to stabilise the segment. However, toe touching works the back hard when it is still quite reactive, and you need to do a lot of spinal rolling to break up the stiff soreness caused by this.

In the final stages of rehabilitation, the emphasis is on spinal strengthening so the weak link does not keep re-inflaming. This is done with horizontal intrinsic exercises off the end of a table. However, the muscular compression of the energetic unfurling action jams the segments together and can make the low back sore again. For this reason, the main intrinsic exercises should only be done once every week to ten days, although it is necessary to continue spinal rolling, rocking the knees and using the BackBlock, to disperse treatment soreness.

Beware! People with spinal instability problems often keep their back sore by doing too many spinal intrinsic exercises (both horizontal and vertical) too frequently and not doing enough 'appeasing' exercise afterwards (knees rocking and spinal rolling). In this regard, lazier patients often do better!

A typical self-treatment for acute instability

Purpose:
Relieve muscle spasm to relieve compression on disc, disperse inflammation of weak link, gap open the spinal segments to introduce

pressure changes, strengthen tummy and switch off over-active erector spinae.

Rocking knees to the chest
(60 seconds)

Reverse curl ups
(five excursions)

REPEAT BOTH EXERCISES 3 TIMES

Medication. Rest in bed. Repeat exercises every 2 hours. See Chapter 7 for information about bed rest and the purpose of these exercises and the correct way to do them.

For how long? Until acute leg pain has gone.

A typical self-treatment for sub-acute instability

Purpose:
Relieve muscle spasm to relieve compression on disc, disperse inflammation of weak link, gap open the spinal segments to introduce pressure changes, strengthen tummy and switch off over-active erector spinae, break up spinal impaction from muscle spasm, re-establish intrinsic muscles' segmental control, and decompress spine to promote disc regeneration.

Rocking knees to the chest (60 seconds)

Rolling along the spine (30 seconds)

Reverse curl ups (15 excursions)

REPEAT THE THREE EXERCISES 3 TIMES

Segmental bridging
(up and down 2 times)

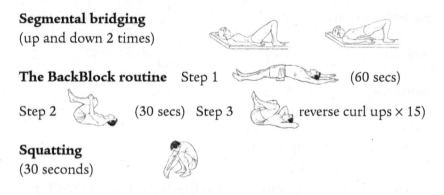

The BackBlock routine   Step 1   (60 secs)

Step 2   (30 secs)   Step 3   reverse curl ups × 15)

Squatting
(30 seconds)

REPEAT FINAL THREE EXERCISES TWICE

    Repeat regimen every evening. If the leg pain returns, revert to acute regimen for several days, or until it ceases.

    For how long? Until intermittent leg pain has completely gone.

A typical self-treatment for chronic instability

Purpose:

Relieve muscle spasm to relieve compression on disc, disperse inflammation of weak link, gap open the spinal segments to introduce pressure changes, strengthen tummy and switch off over-active erector spinae, break up spinal impaction from muscle spasm, decompress spine to promote disc regeneration, and re-establish intrinsic muscles' segmental control.

Rocking knees to the chest
(60 seconds)

Rolling along the spine
(for 30 seconds)

Reverse curl ups
(15 excursions)

BackBlock routine   Step 1  (60 secs)

Step 2   (30 secs)   Step 3   (15 times)

Squatting
(30 seconds)

REPEAT ALL FIVE EXERCISES 3 TIMES

If you *do not* have leg pain do the following exercises ONCE per week/10 days

Spinal intrinsics strengthening
(5–10 excursions)

Rolling along the spine
(30 seconds)

Repeat regimen three times per week, except for the intrinsics exercises which should only be done once per week to ten days. Expect the back to feel worse after intrinsics exercises but if it remains so for more than a couple of days, cut down to one session every two to three weeks.

7

Treating your own back

This chapter discusses all the procedures involved in treating your own back. It explains the rationale behind each one and what it will do for you. It also spells out the do's and don'ts of each procedure, and the common pitfalls of each one.

HELPING YOURSELF

With self-treatment, it goes without saying that there is no physical input from anyone else. Although I strongly suggest that in the beginning you see a physiotherapist, manual therapist, chiropractor or osteopath who will isolate your problem and initiate the un-jamming process, the rest is up to you. There are many operators who are expert at undoing complex jamming of vertebral segments, but they alone cannot deliver you from trouble. You need to help yourself and be confident about doing so—and being effective! Even the best therapeutic magician, who plays the spine with the finesse of a concert pianist, can only address the 'hands-on' aspect of your care. The background decompression of your spine and the restoration of your trunk control can only be done by you.

Spinal therapies which do not recruit the sufferer enjoy short-lived success. Yet most people would love to help fix their problem, if only they knew how. A quick tweak here and another tweak there, and a repeat appointment in a couple of weeks, rarely achieves anything if you have not done the groundwork in the meantime. The regenerative period may be a fraction of the degenerative one, yet it still takes time. It is a journey which must methodically unfold as it stops the destruction and turns it around. A condition which has taken years to evolve cannot be cured in a moment, especially when the factors which brought it about (gravity and our upright stance) never go away.

You have to decompress your spine. No-one else can do this for you. You also may have to loosen contracture of the spine's soft tissues. No-one else can do that for you. You have to restore strength to weak muscles. No-one can do this for you either. So except for the subtle art of manually loosening a spinal segment, which is hard for you to execute with any degree of accuracy, you do the vast bulk of rehabilitation yourself. And remember, you have the invaluable advantages of intuition and instinct guiding you from the inside.

The fundamentals of self-treatment are to minimise the compression and restore elasticity to an immobile link. Restoring segmental freedom introduces mobility which allows the disc to undergo pressure changes to bolster nutrient traffic and stimulate cellular metabolism. As a consequence, the disc can suck through and hold more water which improves its health and also makes it more resilient. Thereafter, the disc can also absorb shock better and is less victimised by trauma. A properly hydrated disc also spares the facet surfaces excessive contact. It acts like a pivot on which the segment tips, while the intrinsic muscles at the back control the forward nodding of the vertebra like horse's reins—all of them working at their most efficient angle of pull. When a disc flattens, there is less of this see-sawing action, and everything works less well. Strain and eventually pain sets in. Gradual restoration of disc height is thus the first objective of treatment. The key to the remedy is as simple as knowing the cause.

Preliminary thoughts

Self-treatment of a spinal problem involves graduated combinations of a few simple exercises, rather than a great variety. And because the central jamming is the first disorder from which other conditions flow, the fundamentals of all treatments are just the same—even for more complex disorders. Whatever the problem, the same few exercises keep cropping up: rocking the knees to the chest, spinal rolling, squatting, using the BackBlock, abdominal strengthening exercises (reverse curl ups) and intrinsic spinal strengthening. But before starting these, it is important to know that treatment of any problem must be taken at the right rate. In what is essentially a healing process, the regimen must be embarked upon with determination, subtlety and intuition. There should be a fine balance between rigour and rest. It has to mean business but cannot be rushed. You have to do what is necessary, unflinchingly, but you cannot harry your back. You must tailor treatment to your spine's ability to recuperate after each 'new' activity.

Be guided by your instincts. You do have to keep pushing the spine at times when you might be fearful about it but remember, most pains are good pains and most people are fearful of the wrong wrongs. They are too defensive and too ready to brace against the pain, wrapping it up, clenching around it and keeping it inside the spine.

Incidentally, it is surprising how dramatically you can reduce your discomfort by willing the muscles around your pain to relax. You can do it during any activity: walking along, waiting for a lift, or stretching to make a bed. As you feel the muscles start to grab, concentrate on making them let go, like a meaningful glance to stop a naughty dog jumping up. Subtle—even bizarre—as it sounds, a simple breakthrough like this is a huge milestone in the management of your back problem.

When self-treatment fails, it is often because you have been too tense to disengage from the pain. At the same time, there may have been too little (or too much) physical effort. You may not have been sufficiently calm and persevering or you may have been too vigorous, with too high an expectation of an instant cure. Alternatively, it may have been progressed too recklessly. The clue with self-treatment is to keep going, quietly yet purposefully: not too timid, not too aggressive. Just keep going with thoughtfulness but not obsessive introspection.

Chances are, you will occasionally lose control of the treatment and things will seem to go backwards. You will have a hiccup in progress when everything seemed to be going swimmingly; a savage twinge as if your back has turned around and bitten you. When this happens, you will be cast into despair and lose your serene overview, and your confidence will waver. More importantly, it will stop you dead. You will fear the worst and you will be too frightened to keep moving forwards. But if you grind to a halt, the back will have taken over again.

Rest assured that at some stage, *all* of you will have to deal with some kind of wobble in the path of recovery. You will think you are getting nowhere; that it shouldn't be this painful. You may feel weak or nauseated or sorer than yesterday. There may be newer pains higher up, or a different type of pain in the same area of your back.

Keep calm and ride out the storm. Employ all the resources you have to avoid panic. Your back is simply complaining about the new rules, and it is very important not to succumb. You need to ease up for a day or so, *but do not stop*. Remember you are on a one-way trip, and the direction is forward. Having stirred up the very centre of things, it is critical to keep going so that in the end there is something to show for it, after the reaction has died down.

You will not reach this point if you abort part-way through. You will be left with the sense that whatever you did made you worse. You will have provoked the root cause into an angry reaction lasting several weeks, even months, without the follow-through to see any appreciable gain at the end.

THE PROCEDURES

Bed rest

Sometimes a back problem is too inflamed for exercises to start. When this is the case, the ideal approach is to rest in bed with medication, performing gentle bouts of exercise throughout the day.

Bed rest simply means going to bed, which is usually hard to do. The adjustment in peoples' lives is never easy and invariably you think there must be a quicker way. But to do it properly you simply must rest. Furthermore, you have to give yourself over to the notion of doing nothing, with an attitude of abandon rather than frustration, otherwise you will get nowhere. Both physically and emotionally, you have to 'let go'.

With cases of severe inflammation, there can be a build-up of treatment soreness which mimics the original pain. If the condition is chronic, the increase in pain is tolerable, but if the segment is acutely inflamed, it will seem as if the treatment has made you worse. Bed rest lets the dust settle. The peaceful periods between the hard work let the inflammation subside and allow your back to recuperate so therapy can go on.

Bed rest also eliminates the compression of gravity, which, as a first measure, starts the un-jamming of the spinal link. This first step usually makes the worst of the pain start to fade. Apart from relieving the local engorgement of the problem link, the horizontal resting eases the spasm of the spinal muscles. As they relax, the pain eases as the circulation picks up. There is better natural sluicing of toxic inflammatory products away from the local nerve endings and the pain is no longer ever-present. As the protective spasm eases, the restorative process gathers pace.

The correct way to rest in bed

The important thing about bed rest is that it must be horizontal. It is not as effective on the sofa. You can get up from your bed to shower and get dressed but you must go straight back to bed and stay there, perhaps for several days.

Use a pillow only under your head, two at the most. Avoid using a stack of pillows so that your back hangs in a deep slump. If you are very uncomfortable even lying there, you should have one pillow under your head and several pillows under your lower legs so that the hips and knees are as close to a right angle as possible. This reduces the pressure on your low back.

While you are lying on the bed, try to keep active. Don't lie there rigid because that is the opposite of what you are trying to achieve. Keep relaxed and mobile, always flat, filling your time with the exercises shown below. Rest any way you are comfortable (although all positions become uncomfortable after too long and you will have to move). Gather all your things about you: telephone, books and chairs where people can sit. Your back needs time and peace, so sign off and enjoy the rest.

Take care when getting up. You will have to roll yourself over to the edge of the bed and swing your legs over the side. To lever yourself up sideways, you will have to push into the bed with both arms, with your tummy braced. Your legs will go down to the floor as your trunk comes up. When doing bed rest properly you should only get up two to three times per day.

Do remember though, that getting up by rolling over sideways like this should only be part of your regimen when your back is at the height of crisis. As soon as possible, you have to get used to getting up out of bed normally, by rolling forward and sitting up properly. Continuing to roll over like a log is not helpful once you are on the mend because it keeps you in the territory of an invalid. In the same way as bending with the back jacked straight and sitting up too straight in a chair keep your back bad, getting up with a clumsy robotic action keeps the spine working like a rigid column rather than a segmentally undulating one.

Medication

Taking tablets is as unpopular as bed rest. People are wary of entering an arrangement which may have no limits. But in the way that going to bed interrupts the chaos of running a life through pain, taking medication can allow you the space to regroup and soften your mind-set about your back as well as lessening the pain. By surrendering to the very foreignness of taking tablets you may also free yourself from all your previous, rigidly held dictums. Medication, particularly muscle relaxants, often helps wipe the slate clean to set a new course in the self-management of your problem. By allowing a drug-induced reprieve from feeling your pain and thinking about your problem (which is often more important), medication lays the groundwork for the calm and focused dawning of a new era. Appropriately prescribed medication, in combination with physical influences, can be just what is needed to get you over the worst.

The three different categories of medication are painkillers, muscle relaxants and non-steroidal anti-inflammatory drugs (NSAIDs) and they must all be prescribed by your medical practitioner.

Painkillers and NSAIDs

The choice of painkillers and anti-inflammatory drugs needs to be discussed with your doctor. He or she will know from your medical history if there are contra-indications to taking them and will be familiar with features of the different drugs. Getting rid of pain is the main objective so the more powerful the painkiller the better, but it should be for a limited period only and you will need your doctor to supervise. Tablets should be taken three times per day—morning, noon and night—so that the pain is kept at bay at all times. Remember, pain makes pain. The less pain there is, the less the body reacts to it. (There are many painkillers on the market but be aware that ones with a codeine base cause constipation which can make back pain worse.)

NSAIDs also come in many different forms with brand names like Naprosyn, Voltaren, etc. Their role is to target and actively quell the inflammatory process which is the primary source of pain. Under the blanket cover of less pain, they allow the part to get working normally, but more importantly, they allow more vigorous treatment. NSAIDs should not be taken flippantly. They irritate the bowel and cause nausea, but they must be taken consistently over a set period because their effectiveness depends on maintaining a certain level in the blood. They should always be taken with food to reduce their noxious effects.

In a sense, treatment irritates the tissues. It provokes an inflammatory response from all the tugging and stretching to get a segment going again, which brings it to the very brink of what it can take. Treatment is designed to provoke a reaction which brings the blood rushing but, in doing this, it walks a tightrope between coaxing the body along at a rate it can accept, and overwhelming it. Treatment amounts to 'tailored movement' with all the knockabout excesses of everyday activity smoothed out. But it can also swamp the joint, especially if it is semi-immersed with inflammation to start with.

If you are not careful you can develop an acute treatment reaction on top of the underlying chronic one; additional 'man-made' inflammation with more pain on top of the pre-existing lot.

Sometimes the reaction can be so severe, it feels as if the treatment has made you worse. Usually even the worst reactions leave you better off in the long run, although this is often hard to believe at the time. It takes some convincing that all the 'new' pain has anything going for it at all. Provided you do not lock up too much with muscle spasm when the going gets rough (which *can* cause you to be worse off afterwards), the degree of treatment reaction is usually directly proportional to the amount of gain. Furthermore, it actually *helps* at the time to see all the soreness and newfound fragility in your back as a good sign. It helps too when you recognise (which you often can, if you keep clear-headed enough) that the treatment reaction has a different quality to it, so the 'new' pain is actually different from the 'old' one. All these are good signs, although the discomfort nevertheless has to be worked through.

Even so, it is better still to minimise the pain at the start; to knock the peak of it off at the source, to stop the cycle gearing up. Better to take the medication in the expectation that treatment will almost certainly cause a reaction. Better to keep it under wraps, so that when you emerge from the end of treatment with proper function restored, it is like a fog lifting. You stop the medication as the pain fades, and walk free from the rigours of treatment and the anxiety of taking pills.

Muscle relaxants

For various reasons the muscles can take over and make things worse. Sometimes fear alone is sufficient to do this, but at other times the inflammation is so intense, the muscles go into protective mode as part of the spiralling reaction. Muscle spasm is a normal phenomenon when you have pain. However, the degree of spasm for a given level of joint irritability varies greatly from one person to the next. It depends on many factors, not least personality type and the presence or absence of other emotional stresses in one's life—some of which may be buried in the subconscious.

It is no overstatement to say that muscle spasm alone can transform a nuisance problem into a nightmare. If it takes hold it can cause things to slide so far that the back can become incurable. Muscle spasm is the wild card with backs. Its bad feature is that it may keep on keeping on, long after the original cause for it has gone away. In other words, it can be the sole stimulus which keeps the inflammatory cycle going. The clenched muscles are painful in themselves (as is any muscle suffering low-grade cramp) but their continuous clench stops blood passing freely through the part. There is pain because the muscle fibres are working overtime and because there is not enough oxygen for their needs.

Muscle spasm is often interwoven with anxiety and depression about the problem which can be enough to keep the cycle going—and it is for this reason that muscle relaxants have a role to play. The easing of the muscle hold, even though chemically induced, can break the nexus between pain and reaction-to-pain and create a vital breathing space for recovery.

Muscle relaxants ease muscle spasm, whether that spasm is caused by fear or has an organic origin. Diazepam, otherwise known as Valium, is the best, although it is not without detractors. When your back cannot admit any local movement without a nasty growl of pain, Valium dissolves the grip of the muscles (although it makes you quite stupid in the head). Valium is both addictive and cumulative and needs to be taken under the strict guidance of your doctor so it doesn't become a problem in itself. Its best use is with the first glimmer of an old pain returning. If you have ricked your back and it feels stiff with the first signs of pain down the leg, one Valium and an early night's sleep is often enough to see it off.

At the height of an acute episode the dose should be high enough to cause drowsiness. The ideal amount is 5 mg three times a day (one pill on rising, one at midday and one after dinner at night). It should make you want to be in bed, docile and floppy like a rag doll so your spine can ease apart at its painful crimped links. It should be enough that when the episode has passed, you cannot recall the sequence of events or the days passing. It can be tailed off once the mobilising is well in train and the initial treatment pain has passed.

EXERCISES FOR TREATING A BAD BACK

Once the stage is set with adequate rest and any necessary drug therapy, the following exercises reverse the structural and physiological changes of the motion segment. Basic theoretical treatment regimens, using various combinations of the exercises for specific back disorders, are in Chapters 2 to 6.

Rocking the knees to your chest

This is the ultimate 'appeasing' exercise. It is the least taxing and therefore least frightening exercise in the early loosening of a jammed segment. It is performed in the horizontal supine position to eliminate compression of the spine.

The primary function of the knees-rocking exercise is to fan open the posterior compartments of the spine like flaring out a deck of cards. The action stretches the muscles down the back of your spine when their tonic hold has pulled the interspaces shut. Releasing the muscles un-jams the facets and releases the pincer effect on the intervertebral disc. Thus the passive stretching inhibits the additional closing down effect across the inflamed vertebral segment. By providing 'active' decompression it produces the first glimmer of the spine lifting off the compressed segments. It is a very efficient first-base exercise.

Like spinal rolling which you will read about below, rocking the knees to the chest is very effective if you have just jarred your spine or hurt it in some way. Rocking has an immediate neurophysiological 'switching off' effect which defuses the alarm and preempts the local muscles locking up the spine. The to-and-fro rocking action familiarises your back with movement again so it doesn't have time to get stiff. It keeps your back moving in a non-threatening way and encourages the fundamental physiologies—active disc metabolism, unhindered blood flow and proper drainage—to resume. Disc nutrition is enhanced by incremental amounts of fluid being pumped through, while the pressure changes stimulate the synthesis of proteoglycans which thereafter provides a more active osmotic pump.

In the acute phase the pull-and-release evacuates the inflammatory products from the site of injury. With more chronic conditions, the main benefit is the physical stretching of the tight structures. The non-weight-bearing loosening of all the soft tissues of the posterior compartment, including the back wall of the disc, immediately gives the tight segment more freedom to move. The much vaunted breakthrough of injecting one's own healthy disc cells (autologous) into degenerated discs is doomed to failure if the target disc is not made to 'suck and blow' after the procedure has been done. If it remains static it is impossible to keep the new cells alive.

Although rocking the knees to the chest is most effective for segmental stiffness of L5 and L4, it is also effective higher up in the lumbar spine where there may be some rotation of the segment as well. It is particularly effective at breaking up what seems like armour-plated stiffness across the back of your pelvis that often accompanies an acutely sore back. You can move your legs carefully by manoeuvring your thighs as levers to drive around the back of the pelvis, feeling for hard patches—and this is where you tarry. You can minutely oscillate the thighs up and down, left to right from dimple to dimple, or you can trace a circular outline around the apices of your triangular-shaped sacrum, first one way then the other. As you get more adept, you will feel that the hard patches coincide with the painful spots and this is where you stay, rocking gently on them until they 'melt' and the going gets easier.

Simple as it sounds, rocking the knees is often difficult to get started, let alone do well. If your back is in acute distress it is not easy to get the knees to the chest. Your legs feel heavy and your spine is loath to bend, and you may get stabs of pain as you grapple with lifting your thighs. As the spine starts to round more easily, your hip joints often complain about being bunched onto the chest. It may be more comfortable for the hips if you allow your knees to part widely around the abdomen. (You may also find that one leg is more comfortable doing this movement than the other.)

With the more chronic disorders, where your lower back may have been stiff for decades, the cement-like rigidity is often unwilling to yield. The lower segments move as a rigid block, with all the hinging taking place at the hips and the upper lumbar levels. As the segments ease apart, your back starts to round more easily, making it a natural progression to spinal rolling, which is discussed below.

The emphasis with rocking the knees is to keep the movement as subtle as possible with the arc of movement only a few centimetres. Don't be tempted to tug at your knees with the muscles in your neck standing up. Don't jab your head up to meet the knees. Leave your head and shoulders calm and relaxed on the floor and gently oscillate your legs with the fingers interlaced around both knees. As your back relaxes, a sense of movement will dawn, like a piece of frozen meat thawing. The tightness will fade as you feel the vertebrae gapping open at the back. Don't hurry and try to imagine the vertebrae segments pulling apart.

The correct way of rocking the knees to the chest

- Lie on your back on the bed or on a soft surface on the floor.
- Brace your low back by sucking your navel in hard. If your tummy is weak, push in with one hand on the front of the tummy for reinforcement.
- With the other hand behind your thigh, pull one leg up to the chest. As soon as one foot is off the bed you can use both hands to pull the leg onto the chest.
- Bracing your tummy in the same way, pull the other leg onto your chest. Crossing your ankles makes the legs less unwieldy.
- Cup a hand over each knee and then move your legs so the thighs rest at 90°. Oscillate gently in this position, with the movement almost imperceptible.
- For high lumbar problems, take your knees closer in to your chest.

If your hips are tight you can cross your ankles and link your wrists across your upper shins and oscillate in this position.

Rolling along the spine

Spinal rolling seems so easy yet it is very sophisticated in a neuro-physiological sense. The alternate tipping back and forth along the spine helps break up the domination of the over-active erector spinae muscles and the corresponding reflex inhibition of your tummy muscles. As you tip towards your toes you use the abdominal muscles and as you go back the other way you use your back muscles. By 'selectively recruiting' both groups you interrupt the status quo when your back is bad: when your back muscles will not switch off and your abdominal muscles cannot switch on.

Spinal rolling is very important in self-treatment. It breaks up the superficial brittleness of the column and is the simplest form of spinal mobilisation. Each vertebra has its turn at gliding past its neighbours as your bodyweight rolls over it. Again using the analogy of the keyboard, rolling along the spine is like rolling your forearm and wrist down over an expanse of piano keys as they depress one after the other. Each vertebra has its own musical note of stiffness, although the jammed one is especially shrill as you press over it.

In this regard, rolling along the spine is a primitive diagnostic tool. By rolling over the segments you can isolate your problem level. You can 'examine' the L4-L5 and the lumbo-sacral interspaces by clasping your hands behind your knees and nearly straightening your legs with toes high in the air. The weight of the legs and the long leverage make it easy for you to tip back and forth over the lower end of the spine to see whether you elicit pain. To isolate the mid-lumbar level, hug your knees a little closer to the chest so that they make a shorter lever. Depending on the bulk of the upper chest, the angle at the knees will be closer to a right angle. To isolate the thoraco-lumbar level, you have to shorten the lever even more by having your legs pointing almost straight up to the ceiling. To tip your weight towards the higher end of the lumbar spine, you simply bend and straighten your knees in the air, which rocks the body back and forth.

All these movements require a fair degree of control and will not be possible in phases of acute pain. Simple as they sound, it is often too painful to get your spine rounded sufficiently and you struggle about stranded, like a beetle on its back.

Or it may not be painful at all, simply stiff. The patch of immobility may be loath to press out the other way because the vertebrae cannot glide backwards. (I can see this as a small hollow scoop in the low back when you bend forward from the standing position.) It makes rolling along the spine like bumping over a square wheel, with a clonk as the flat patch hits the floor. It requires extra pulling in below the belly button to shrink in the lower abdomen, to force the stiff patch out the other way.

As a therapeutic exercise, spinal rolling is the all-round panacea. It is effective first thing in the day if there is early morning stiffness. It dissolves guarding muscle spasm which can hold your back as rigid as a plank. It is also useful if the spine has just been jarred; in this case, the rolling should be as relaxing as possible, along the whole length of your spine. As everything loosens up, your legs tip right back over your head and as the spine softens it is easier to isolate the problem level.

Loosening the specific vertebra requires small range pivoting, back and forth with small amplitude excursions, right on the painful spot. You have to grab your knees and steer yourself back and forth with precision. Working the vertebra free is like pressing out the stiffness, pivoting back and forth on the carpet. You can also pause in mid-flight, staying motionless on the spot to allow the spine time to sink down and relax around it.

Note: Don't be confused by the picture following. Most times, this exercise will be performed as small 'pivoting' spinal rolls, almost on the spot. This prises the specific problem vertebra free and settles local spasm.

The correct way to do rolling along the spine

- Fold a bath towel double and place it on a carpeted area of floor to roll on. Do not attempt to roll in bed.
- Lower yourself down carefully to lie on your back on the floor.
- Gather up both thighs and link your fingers under them, or around your knees, whichever is easier.
- Lift up your head and neck so your low back makes a wide rounded 'U' shape on the floor. (To help keep your upper trunk forward you may need to bend both elbows out to the side as you hold your legs.) The stiffer you are the more difficult it will be to get into this position.
- Once in position, rock gently back and forth along the spine with small amplitude movements.
- Attempt to pivot on the stiff link in your spine which will be obvious by its 'bruised-bone' tenderness as you roll over it.
- Use your legs for leverage. As you straighten them out they alter your weight distribution and make it easier to tip towards the lower end of the spine. If you keep your legs bunched up on your chest it will focus the rolling towards the upper end of the spine and require more jerking effort of your head to bring it down to the lower end.
- Continue for 15–30 seconds, trying to relax as much as possible as you do it. Let the gentle rocking motion mesmerise you.
- To rest, put one foot on the floor, and then the other, holding on with your tummy as you lower each leg. Leave both knees crooked, feet flat on the floor.
- Resume at one-minute intervals and repeat up to three times.
- This is not an easy exercise to overdo. Cramp of the tummy muscles and the muscles at the front of your neck may be the only things that stop you.

Legs passing

This exercise is an effective way of tricking the tummy muscles to switch on when they reflexly inhibit through the presence of back pain. It encourages the back to let go and the hips to swing freely with walking, which they do not do when your back is painful. It also encourages the abdominal muscles to hold the pillar of lumbar segments secure as the legs move and this protects the low back from ongoing micro-trauma during everyday life. It is a very important first exercise for getting the tummy muscles working. As you do it, make a mental note of the degree to which your tummy muscles are working to hold the midriff stable and attempt to replicate this when you are up and walking about. *This is how your tummy should work when walking to help off-load the spine!*

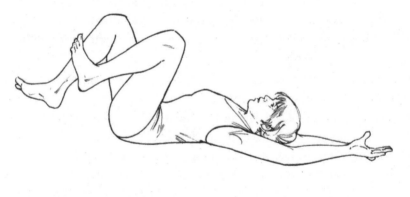

The correct way to do legs passing

- Lie on your back on a soft carpeted floor with your head in your arms.
- Tighten your tummy and press your low back into the floor to remove any lumbar arching.
- Bring your left knee towards your left armpit with the knee fully bent.
- As the left leg starts its return to the floor, lift the right leg so the legs pass in mid air, at all times keeping the low back pressed into the floor.
- This is not a 'bicycling action' so you do not straighten either knee at any stage as this strains your low back.
- As each leg comes down, just brush the floor with the sole of your foot between excursions—that is, do not fully rest your leg on the floor.
- You will feel the midriff fully switched on as you control the weighty legs alternately passing one another.
- Continue for 60 seconds, moving the legs slowly and with control.

Reverse curl ups

Reverse curl ups strengthen the abdominal (tummy) muscles and are preferable to old-fashioned 'sit ups' because they do not involve sitting on the pelvis. They also reduce the likelihood of shearing strain across the lumbar segments, as well as exerting less pressure on the inflamed lumbar segment. Although sit ups, spinal crunches and a great variety of abdominal exercises are popular with gymnasiums, they over-work the upper abdominals and tend to lead to spinal compression. On the other hand, reverse curl ups specifically recruit the lower abdominals and in most respects are a less problematical way of strengthening the tummy. Reverse curl ups are relatively strenuous compared to normal sit ups and they have the advantage of demanding greater participation of transversus abdominus. High lumbar problems (from L3 and above) also do better with tummy strengthening done as reverse curl ups.

They should always be done lying on the back, using the abdominal muscles to pull the knees back and forth towards the chin from a starting position of 90°, lifting the bottom as high as possible off the floor. The emphasis should be on drawing in the tummy and bringing knees to the chin rather than chin to the knees. Never lower the legs beyond 90°, nor start the movement by lifting the feet from the floor, and returning them there; the weight of the legs can strain the back.

Try not to jerk the legs up and then let them go. Make sure that both excursions take the same time as you lower your legs under control. If you are weak, the best way to incorporate transversus abdominus is on the return journey. Keep your back round, and press the whole lumbar area into the floor, lowering one cog at a time. Drawing up the pelvic floor also helps. If your back clicks or hurts as you do the movement, pull your tummy in harder and reinforce it with one hand pressed in, rounding your lower back more.

Tummy muscles play an important role in off-loading the base of the spine. They do this by co-contracting with the back muscles which tenses the abdominal cylinder and raises the abdominal pressure. This lifts the spine skywards and increases the tension between the vertebral links. Strong tummy muscles also control the way the vertebrae move during upright activity. They help the vertebrae to tip forward rather than shear when the spine bends. Thus tummy strength plays a critical role in optimal spinal performance.

For this reason, reverse curl ups are probably the single most important exercise in the prevention and management of spinal problems. In short, they not only prevent the spine harming itself but, once a spine *is* in trouble, they help get the segments back to full function.

When a low back is painful, abdominal strengthening plays an important neuro-physiological role in reducing the hold of the back muscles. When worked strongly they automatically make the long back muscles switch off. This simple-is-genius truism of physiology is a very effective way of making a stiff back relax. Reverse curl ups open up the spinal segments at the back while the tummy shrinks in at the front. The elongation of your spine stretches and softens their muscular clench and allows the spinal segments to ease apart.

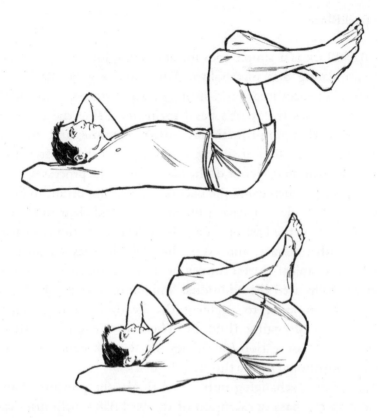

The correct way to do reverse curl ups

- Lie on your back on the floor and take first one knee to your chest, then the other.
- Crossing the feet at the ankles and letting the knees fall apart comfortably, take the thighs to 90°, and interlace your fingers behind your neck.
- By lifting your bottom clear of the floor use your tummy to bring your knees up under your chin.
- As you relax and your legs come down, make sure they do not pass beyond 90° which would cause your back to arch.
- When you get into your stride, you can swing your knees up with some gusto, as long as you do not jar your back.
- Repeat fifteen times.

The BackBlock

The BackBlock is a simple tool for alleviating lumbar compression and stimulating disc metabolism. You could say the BackBlock's main role is to accentuate anti-sitting postures since it is sitting that most compresses the spine's basal segments. Using the BackBlock emphasises this non-functional 'other' extreme and foils the sustained compression of sitting. It also invokes the negative forces required to stimulate vigour of the discs' metabolic processes.

At all times, when using the BackBlock, it is important to bear in mind that discs do not have a blood supply and their nutrition is tenuous, even at the best of times. Discs degenerate much earlier in life than other tissues, and even when healthy may be unable to survive mechanical duress under adverse nutritional conditions. Discs critically need the additional 'imbibition pump' which phys-ically sucks in water to augment the steady 'osmotic pump' of proteoglycans dragging fluid in. These two engines—diffusion (osmosis) and convection (pumping)—must work hand in glove to conduct nutritional traffic through the discs.

Apart from exchanging nutritional fluids, the pressure changes induced by the passive extension of the BackBlock help discs carry out essential *maintenance and repair*. The disc cells most active in manufacturing proteoglycans are directly stimulated by variations in disc hydration; thus a vigorous tidal exchange actually stimulates regeneration of the very cells that help attract water in the first place. The opposite is also true as discs degenerate: the more meagre the fluid exchange, the more sluggish the multiplication of proteo-glycans, the magical x-factor of discs that attracts and holds water. The discs slowly starve as their increasing immobility locks out movement and the regenerative cycle slows down.

In addition to the pressure-induced physiological processes, the direct physical stretching of stiffened disc walls makes them more compliant and accommodating to imbibed water. As more fluid is retained in the nucleus, this part of the disc can shoulder load again, thus alleviating compression of the sensitive outer disc wall.

In this way the typical pain of a 'stiff spinal segment' is lessened. With better clearance between adjacent vertebrae as disc height is restored, there is more room to move between the bony parts of the spine and the typical pain of 'facet joint arthropathy' is also alleviated. In cases of 'segmental instability' where reduced hydrostatic pressure makes the disc a weaker spinal connector, augmenting the water volume of the drier disc can make the segment more secure.

Apart from pulling the segments apart the BackBlock re-establishes the spine's proper 'S' bends. If the back is too stooped it straightens it. In the thoracic area it opens out a round-shouldered look, and at the keystone of the spine—the base—it pushes the bottom in and straightens the entire skeleton. On the other hand, if the low back is too hollow, it can reverse that too. Lying passively over the BackBlock, the vertebrae inch backwards and the high arch of the lumbar area slowly drops flatter towards the floor.

The BackBlock is the natural antidote to our 'C' shape of habitual slumped sitting. Placed under the sacrum and the thorax in turn, it prises out the hours of compression and stretches the soft tissue contracture of the front of your spine and hips (which have also spent too much time bent). Thus the BackBlock uses gravity to press your skeleton out straighter, aligning it more trimly over its narrow base. In this way it deals with its two main pain-making causes at once: basal compression and the spine's dynamic imbalance over the sacrum.

Although our nearest cousins the apes have a rounded 'C'-shaped stature, with their heads carried well forward, in front of the line of gravity, they can reverse any problems this might cause by swinging through the trees on their arms (brachiating). Human beings have no such means of everyday traction to straighten their spines, and must counter the ill-effects all their lives. Most of us have to work at keeping a good posture and decompressing our spinal bases other-wise we are all in line to develop low-back pain. A BackBlock—nothing more than a simple chunk of wood (or plastic which is lighter for travelling)—is a most effective tool.

You must always start off using the BackBlock on its flattest side. As your legs drop down and you feel the pelvis pulling off the spine's base you often feel a tug in your back, right where it hurts. It feels as if it really means business; as if a screwdriver is drilling right into your pain. At first, it may take some concentration to relax and let the discomfort fade. Some people cannot even lie on their back on the floor, while others must make do with a lesser form using a flattish book or folded towel under their sacrum. It may take several weeks to progress to using the BackBlock. Usually, you progress from the flattest side to the middle height within a couple of weeks, while others go straight to it if they feel nothing is happening. You should not attempt to use the BackBlock on end first off.

Only remain on the BackBlock for one minute (60 seconds), before following with the mandatory steps 2 and 3. Remember, greatest fluid shift through the discs happens at the point of loading and unloading, and that it is pressure *changes* you want. So it is better to do repeated short sessions rather than one long one. Also, if you stay there too long it can be difficult to get off. Even so, there is always a murmur of discomfort as you go to lift your bottom off the Block.

If you do stay on the BackBlock too long, what should be a little pain becomes a big pain as you go to lift off. Although this does no harm, it can be unnerving and you will think the BackBlock does not agree with you. Better to do shorter sessions, followed by the usual rocking the knees to the chest and reverse curls after the BackBlock has been removed.

It is imperative to *always* follow the BackBlock with the low abdominal exercise of reverse curl ups. After rocking the knees has first removed the castness of your back, and accommodated it to humping round the other way, you should go straight to them. If you fail to do the prescribed number (or worse still none at all) your back will feel very stiff and sore over the next few days. Reverse curl ups and the BackBlock should always be balanced, otherwise the benefit is reduced. If people are sore after using the Block, they have invariably forgotten their tummy exercises.

The BackBlock under your upper back usually creates a greater feeling of well-being, there and then. Apart from taking the hunch out of your spine it makes you feel taller and looser in your own skin, especially when you take your arms over your head, inter-linking fingers and turning the palms away. You can feel your whole spine pulling out of your pelvis like a cobra out of its basket with a beautiful emancipating stretch which goes right from your chest through to your waist and low back.

The BackBlock should be placed on its flattest side, lengthways under the long curve of the thoracic spine, so the top edge is level with the top of your shoulders and the lower edge at high waist level (sometimes, if you are short, it digs in here until you loosen up over the next couple of days). As you lie back, you may need to lower your head back to the floor with your hands and when there, elongate the back of your neck by pulling your chin in. Some people feel nauseated on first putting their head back, but this passes as they get used to it.

To help reverse the pinched-forward look of the shoulders which so often goes with a stooped upper back, you can stretch the pectoral muscles at the front of your chest by taking your arms down to your sides in wide semicircular movements (the 'angels' wings') as you lie over the BackBlock. Chest expansion and breathing control is also enhanced if you do these sweeping movements in time with your respiration. Inhale as the arms go up to above your head, keeping the back of the hands in contact with the carpet all the way around. Take a breath or two at the top and then exhale for the return journey, as your arms come down to your sides again. Both ways, the arms will lift off the floor when they approach the 'eagle stretch' position in the top quadrant of the semicircle. Through both excursions up and down, there will usually be a stretch across the front of the chest and into the upper arms.

To get off the BackBlock when it has been under the thoracic spine you simply roll off to the side like a log. Do not attempt to sit forward because this can strain your neck.

Although it gets progressively less painful as you repeat each one-minute session on the Block, the structural changes take some time to reverse and definitive signs of improvement may be quite slow to be realised. Remember that discs have a very low metabolic rate and that they are slow to regenerate, just as they were slow to degenerate. Over the first week or so the pain feels less heavy and dull and your back works better. A good sign is when the back feels 'lighter', as if you are no longer dragging an armour-plated shell around with you. It feels less dense with movement, as if the segments are freer to move individually, and your hips swing when you walk. Especially in a prematurely 'old' and chronically stiff back, these changes and the easing of pain can feel nothing short of miraculous. With more acute problems however, it is less straightforward. Where there is guarding from the muscle spasm, the BackBlock gives better results if it is taken more slowly, and if you have had a bad flare-up you should stop using it for several days (anything from three days to two weeks) until any newly acquired 'fragility' has dispersed.

Incidentally, you can get the same effect by using a telephone book, or something flatter if you need to. A proper BackBlock is better than a miscellany of books because it can be more easily progressed through its various stages: from its lowest side, when the spine is particularly impacted, to the highest side when the spine is in its final stages of rehabilitation. Apart from its stringent dimensions, it also serves as a neat and timely reminder as you spy it sitting in the corner by the TV. Just having it there makes you more likely to use it. And the BackBlock really comes into its own with travelling. All the carrying of heavy luggage, plus the long hours spent sitting and sleeping on unfamiliar beds means that you often need it most when you are away from home.

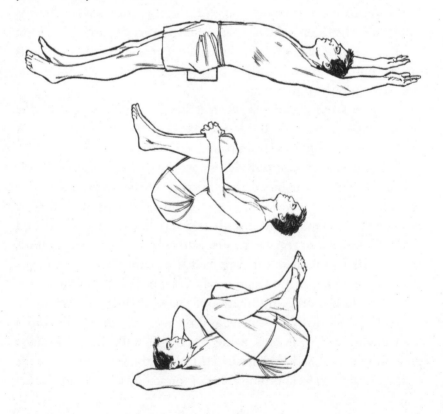

The correct way to use the BackBlock

- Lie on your back on the floor with knees bent.
- Lift your bottom and slide the BackBlock sideways on its flattest side under the sacrum, the broad flat bone at the base of the spine between the two dimples of either buttock.
- Slowly straighten one leg, then the other, by sliding the heel out away from you across the carpet.
- With both legs straight, relax completely and let the weight of your legs pull the spine out. Attempt to keep your heels close together although the feet can roll out. There is frequently a local discomfort at the base of the spine and higher up if another level is jammed there. You should be able to sense the spine pulling apart longitudinally. It is usually an 'agreeable' discomfort, but it should feel as if it means business. Do not be alarmed by the pain. Go with it.
- If you feel absolutely nothing with the BackBlock on its flattest side, then immediately progress to its middle side with the thin edge transversely across the sacrum. You can experiment with sliding the BackBlock up and down under the sacrum fractionally to feel where it is more comfortable. Where it feels best is where it should be. It must *never* be under the spine itself, where it will be very uncomfortable. (I am often surprised how patients forget and put it there.)
- With the BackBlock at the right height, remain in position for one minute, completely relaxed as the legs imperceptibly drop down.
- After a minute, slowly bend one knee and slide the foot up towards the buttock, then the other.
- Lift your bottom off the BackBlock. This can be painful but is no cause for alarm. Move slowly and keep the tummy braced as you slide the Block out to one side.
- Lower your bottom to the floor and then do the rocking knees to the chest exercise. Take knees to your chest one by one, then interlace your fingers around both knees (or cup one hand over each knee—whichever is the most comfortable) leaving your head on the floor. Do not tug at your knees with the muscles in your neck standing up with exertion. Rather, oscillate them gently back and forth, persuasively bringing the spine around the other way into a hump.

Initially this may be uncomfortable, with a sense of tightness across the base, but as you continue you will feel the back rounding and lower interspaces starting to gap open. Persevere until the discomfort across the lower back has eased. This may be immediate or can take several minutes of gentle rocking. When your low back feels more supple, it is time for strengthening the lower abdominal muscles. This is done with the reverse curl ups.

- Do the reverse curl ups in strict accordance with the instructions already given, fifteen in number.
- Repeat the one minute on the BackBlock, the rocking and reverse curls another two times, so that by the end you have done 45.
  This routine is best carried out at the end of the day when the spine is most compressed. Three to four repetitions of the three steps takes between 10 and 15 minutes. A good place to do it is on the floor in front of the television.
- Although the ideal time to use the BackBlock is in the evening, some people have a better day if they do it first thing in the morning. Some of my patients keep another BackBlock at work to use in the lunch hour after sitting scrunched at their desks all morning.
- It is sometimes kinder to do the first round with the BackBlock on the flattest side and then progress to its middle height for the second and third round. It takes several months before most impacted spines are ready to progress to using the BackBlock on end.
- Intensive use of the BackBlock must be entered into slowly. At all times, the period over the BackBlock must be matched with equal time doing reverse curl ups. If you use the Block without these afterwards your back will be cast and stiff after getting up, and everything will be much more sore and achey.

Segmental pelvic bridging

This exercise is very effective for restoring segmental control when the small intrinsic muscles (mainly multifidus) have atrophied due to the spine moving as a block. Apart from being the first spinal strengthening exercise, it also restores spinal pliability by helping the segments move individually. It literally breaks the hold of muscle spasm and allows the segments to jostle freely beside one another and pull apart. Segmental movement and control is essential for returning spinal health. As the segments draw away and compress like a concertina, they suck and squirt fluid which nourishes the discs. At the same time, the induced pressure changes stimulate disc metabolism. Your low back often feels tight and sore after a few repetitions of this exercise, so you should always follow it with the knees-rocking exercise and a number of reverse curl ups.

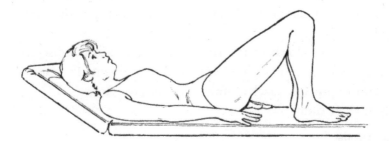

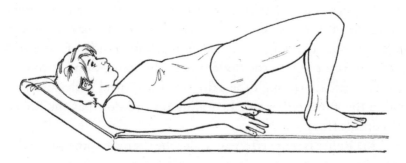

The correct way to do segmental pelvic bridging

- Lie on your back with your knees bent and your feet as close as possible to your bottom and your arms beside you.
- Clench your buttocks and pull your tummy in tight as you tip your pelvis back, making your low back into a round wheel.
- Pressing your low back into the floor, roll up your spine, one cog at a time. Try to feel each segment in turn meeting the floor, particularly the recessed ones which won't press easily into the carpet. Take care not to push your arms into the floor to help you.

- Continue rolling right up your spine to the base of your neck one cog at a time, until your body forms a straight line between shoulders, hips and knees. Rest comfortably, taking weight on the prominent bump at the base of your neck, chin pulled in. For added stretch, you can take your arms over your head, interlacing your fingers and turning the palms away.

- Remain in this position for 15 seconds. *It is important to keep your gluteal muscles switched on by pinning your knees together.*

- Initiate the return journey to the floor by making a horizontal crease across your belly at navel level by sucking your tummy in hard.

- Fold your back down to the floor, one cog at a time, distinctly feeling your spine pass over each spinal segment. It is always difficult to press the back at waist level into the floor.

- Repeat three times.

The Ma Roller

The Ma Roller is an effective way of mobilising the chain of facets running down either side of the spine. It looks like a convoluted rolling pin with two large rounded humps either side of a central depression. You position the Roller under the spine, with the row of knobs you can see through the skin over the central gully, and you mobilise the joints either side by rolling up and down as the Roller moves on the floor.

The Ma Roller is more effective under the thoracic spine where the facet joints are nearer the surface (although it hurts more resting your full weight down) but it is still a useful way of isolating and mobilising stiff lumbar facets. As you oscillate back and forth on the painful spot(s), like a bull rubbing against a low branch of a tree, you feel the bitter sweet pain of the stiff joints being worked.

To specifically target the lumbo-sacral facets you position the Roller just above the two dimples in your low back (you are in the right spot when it hurts more) and hump and hollow your low back in a pelvic rocking action over the Roller. As your back hollows it bends around the prominence of the Roller, letting your bottom reach the floor. As your back muscles relax you can feel the delicious pain of the joints being pushed to their limit of range.

You can get better access to a single joint by removing the Roller and re-positioning one end only under the side of your spine. With the other side of your back lying on the floor you get deeper pressure onto the problem joint, although you must take care not to cause bruising as you seek it out.

In mobilising an acutely tender facet, it is better to use a tennis ball instead of a Roller. The softer, more yielding ball is kinder to the joint and minimises the risk of bruising. A tennis ball is also much easier to carry with you when travelling.

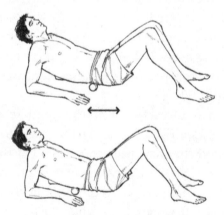

The correct way to use a Ma Roller

- Lie on your back on carpet with your knees bent and feet flat on the floor.
- Lift your bottom and place the Roller under the midline of your back with the knobs of your spine situated over the central gully of the Roller.
- Raising your upper body with your elbows, gently roll your lower spine up and down over the Roller as it makes small oscillations on the floor.

- The problem joints cause a shriller pain as you pass over them. Stay on these points, 'worrying' them as you travel back and forth.
- To avoid bruising, never spend more than 60 seconds working one joint.
- To remove the Roller, lift your bottom up carefully, taking the Roller out to one side.
- Relax the back by lying down gently on the floor and rocking the knees to your chest for 30 seconds.

Squatting

Squatting is the natural predecessor to sitting. In an earlier evolutionary state, although we might have run all day holding a spear, at least we squatted around the campfire in the evenings and eased out the spine's basal compression caused by the day's activity.

Squatting pulls the spinal segments apart and vertically opens the intervertebral spaces from both above and below, rather like extending a tubular fishing net by holding a string at the base with your foot and pulling up the top rim. As the spine elongates, the discs drag apart and suck in fluid.

Squatting comes into its own as the antidote to sitting when circumstances are not right for using the BackBlock. It pulls the segments apart after they have been compressed by the thoracic spine bearing down during lengthy periods perched on the sitting bones. Squatting is the natural partner to the BackBlock. Squatting levers open the back of the disc space more while the BackBlock levers open the front, with the result that the entire circumference of the latticed walls is pulled up.

Deep in the squat your knees often complain, but after a week or so they get to love it and you can stay down longer. Keeping your knees wide while squatting and trying to get as much bend as possible at the hips also helps to release tight gluteal muscles which go with low-grade back problems. The more you can force yourself into the extremes of the stretch, the more you break up the overall picture of tightness and the more easily your spine will bend afterwards.

All of us should squat at regular intervals throughout the day, especially after impact activities like running, walking or playing sport. Sometimes you do it instinctively after standing for a long time, when the spine feels a need to break free of the painful castness of spinal compression.

The correct way to do squatting

- Stand with your heels and toes close together and, holding the side of the bath or a secure rail, bend the knees and drop your bottom to the floor. (You can do it freestanding, as you can see in the picture, but holding on and leaning back gives you a better stretch.)
- Take your heels to the floor and part your knees widely as you take your bottom as close as possible to the floor.
- Bend your elbows to pull yourself forward and drop your head as low as possible between your legs, attempting to turn the full length of the spine into a long, rounded hump.
- In this position gently bounce your bottom to the floor while keeping your head tucked down. Continue for 30 seconds.
- While in this position, suck your tummy in a notch or two and sense the increased separation of the lower segments as the pelvis drops down off the base of the spine.
- To stand up, pull your tummy in and push strongly through the thighs.
- Repeat twice.

If you do not have a suitable object to hang back from as you squat, you will not be able to get your heels to the floor. But you will still be able to sense your spine 'growing' as you pull your tummy in, particularly if you rest your forearms on your thighs to alleviate the weight of your torso. As your back rounds, you can feel your bottom dropping down closer to the floor.

Toe touches in the standing position

This exercise mimics the longitudinal stretch of squatting and also improves the muscular control of the spine. It stretches shrunken facet capsules and back walls of discs (and the other transvertebral structures). Toe touches improve the coordination and strength of all the muscles which control bending.

Although for years people have been told never to bend if they have a bad back, the benefits of doing so are vast. The deep bend pulls the spine out of its vertical clench and releases the discs to imbibe fluid. It also re-educates the power of the intrinsic muscles which specifically control the tipping of the segments.

Toe touches also reduce the over-activity of para-spinal muscles, a common feature of chronically painful backs. They pass control back to the deeper muscles (the spinal intrinsics and transversus abdominus) so the vertebrae are less likely to shear as the spine bends over.

Toe touches have another important role in the later stages of treatment. Done repetitively and with gusto they plump up the lumbar discs. As the spine rhythmically bends and straightens it enhances fluid exchange through the discs, while subjecting them to the pressure changes which stimulate discal regeneration.

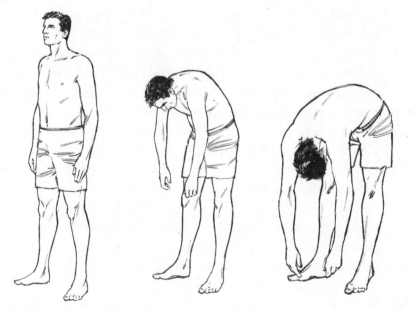

The correct way to do toe touches

- Stand with the feet 15 cm (6 inches) apart and parallel.
- Contract your buttocks and pull your tummy in to shrink the circumference of the lower abdomen, while at the same time drawing up your pelvic floor.
- By tipping your pelvis back slightly so your lower back humps, take your chin onto your chest and curl forward, from the top down, towards the floor.
- Make sure the lower abdomen feels firm and secure, like a tense tube bending with control in the middle.
- Slide both hands down the front of your thighs and on towards the floor.
- If your hamstrings are tight you can bend your knees so you hang there like a gorilla.
- In this position keep the lower abdomen pulled in like a greyhound. Although your tummy will feel pulled in and hard, your lower back should feel a gentle stretching discomfort with a sense of letting go.
- Let your head dangle and your arms flop.

- Try and visualise the bottom vertebrae gapping apart as the bottom of your spine rounds into a hump.
- Without losing the bracing control of the tummy, bounce imperceptibly at the bottom of the bend. The bouncing should be gentle and coaxing, not vigorous. Do several small bounces and then tighten your buttocks in preparation for coming up.
- The return to upright stance has to be done in a controlled way, with the abdomen fully sucked in and the buttocks clenched.
- Come up with an unfurling action, initiated by the pelvis tipping backwards and the rest of the spine following on, the head coming up last.
- In the way that correct curl ups emphasise segmental control, the same is true of toe touches. Especially on the return journey, each vertebra, one after the other, should tip backwards until the upright posture is arrived at.
- Repeat the toe touches three times, trying to get further down with each imperceptible bounce.

Diagonal toe touches

This variation of toe touches comes into its own when treating facet joint problems and chronic disc problems. In the forward bending phase, where you stand with your legs wide apart, taking the left hand down past the right ankle (and vice versa), the capsules of right facet joints are stretched and loosened. You will always find it is harder bringing your hand to the ankle of the same side as your pain. Thus if you have a right-sided facet problem, it will always be more restricted taking your left hand to the right ankle. The inelastic facet capsule on the right finds it difficult to give out and stretch as the tail of the vertebra swings across to the left.

The return journey to upright position, through the diagonal unfurling movement from the floor, is particularly valuable in strengthening the multifidus muscle, usually indicated after a facet locking episode and cases of facet instability. As you come up from taking the left hand past the right ankle, right-sided multifidus pulls on the tail of the vertebra and swings it back towards the midline, thus untwisting the spine. Strengthening this small muscle, which blends so intimately with the facet capsule, helps shore up the joint and makes it less likely to slip out of place. When you have a one-side problem, it is important to repeat the exercise more times to the problem side.

The correct way to do diagonal toe touches

- Stand with your feet 1 metre (3 feet) apart and your hands by your sides.
- Pulling in the tummy hard and clenching the buttocks, bend forward taking the right hand down towards the left ankle.
- If possible, take your hand past the ankle and make small bounces, all the time keeping the tummy pulled in.
- Keeping the hands low, return to the vertical in a diagonal unfurling action, initiated by your buttocks contracting and rolling the pelvis back. Do not take your hands above your head.
- Repeat four times to the problem side to every one to the good side.

Floor twists

This twisting action is done on the floor with the spine unweighted, which puts minimal pressure on the discs as it opens the facet of the upper side. Repeating the stretch in the other direction stretches the disc wall the other way and forcibly closes the facets which had previously been gapped.

The twisting stretch of the spine always makes subsequent longitudinal separation much easier and you will notice this yourself when exercising. The spine feels looser to bend after it has twisted both ways first, and you will always get better separation on the BackBlock doing your twists first.

Floor twists also stretch the nerve root where it has become tethered to the inside of the spinal canal or the intervertebral exit canal. Although root sleeve fibrosis is more common after the inflammation of disc prolapse or facet joint arthropathy, it can also exist with loss of disc height through puckering of the intervertebral tissues and the crowding on the nerve.

To get full stretch on the nerve and its rootlets, you have to get proper straightening of the uppermost knee. The same stretch also increases the tension through the hamstrings muscle of the same side. This is valuable because the hamstrings often retain a low level of contraction when there is chronic inflammation of the low lumbar nerve roots. Apart from being a mild source of pain in itself, the lack of extensibility of the muscle disturbs the sit of the pelvis and shortens the length of stride of that leg during walking. Both factors exert a subtle background influence on the rate of recovery of the spinal problem.

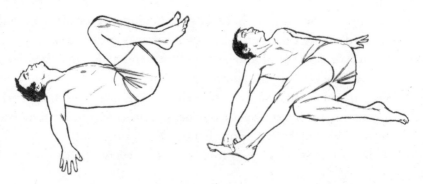

The correct way to do floor twists

- Clear a large uncluttered space on the floor.
- Lie down on your back and bring both knees to the chest, one at a time.
- Place both arms outstretched on the floor at shoulder height with the palms facing the floor.
- Keeping both knees high, let both fall over to the right so the right upper thigh is lying along the floor. As the legs go over, try to prevent the left hand lifting off the floor.
- Make sure that both knees remain as high as possible on the floor with both thighs level, side by side.
- In this position, straighten the top (left) leg at the knee and take hold of the left foot with the right hand. If possible bring the left foot closer to the nose with the right hand.
- Hold for 30 seconds, bouncing the leg minutely by pulling on the toes.
- Repeat to the other side.
- Repeat each way twice.

The Cobra

The Cobra is for stretching—not strengthening. This exercise can be potentially troublesome if it is done too early in rehabilitation. It comes into its own later on to finish the job.

The Cobra involves lying prone with the palms face down on the floor underneath your shoulders. With a straightening action of the elbows, your shoulders are lifted up and the spine drops passively into a hanging arch.

The action does three things: it forcibly stretches the anterior (front) disc wall and the soft tissue structures spanning the front of the intervertebral space; it compresses the back of the disc; it also maximally closes down the facets at the back of the spine which has a regenerative effect on the growth of joint cartilage.

Although the Cobra seems similar to the BackBlock in its effects, it employs a different set of dynamics. It compresses the back of the spine, whereas the BackBlock pulls it apart—a fundamental difference between doing the action prone and supine. The exercise increases the arching of the spine, whereas the BackBlock reduces it (although there is a temporary increase initially until the hip flexors lengthen). The Cobra increases the overriding of the facets at the back of the spine whereas through the backward gliding of the vertebrae, the BackBlock disengages the facet surfaces.

The difference in dynamics is profound and this is exactly why the Cobra is so useful—especially in the later stages of treatment when it is used in unison with the BackBlock.

The extreme arching under pressure 'milks' the posterior (back) compartments and helps reduce the swelling of chronically inflamed facets. It also encourages better fluid movement through the discs. Similar to squeezing the rubber knob on the top of a basting tube, you get better suction up the tube if you expel all the air first. The same is true of a stiff and unresponsive motion segment: if it is pressure-evacuated first, you get better separation of both front and back compartments. At the same time, the pressure dints the cartilage covering the facet joint surfaces. As they un-dint, they suck a fluid exchange through the cartilage bed.

The Cobra also keeps the abdominal muscles long. It keeps them from shortening as a result of the vigorous tummy strengthening regimen. Thus the Cobra is a very effective treatment, as long as the segment is not too irritable.

You can beef up the exercise and improve the coordination of the muscles controlling the spine by switching back and forth between the Cobra and the Pose of the Child. This is where you rest your bottom back on your feet and put your forehead on the floor. As you do this, allow your arms to remain stretched out in front of you, draped across the floor. This increases the longitudinal separation through the length of the spine.

The correct way to do the Cobra

- Lie face down on the floor with your legs straight and your hands on the floor beneath your shoulders.
- By straightening the elbows and pushing on the hands, push your upper trunk back off the floor.
- Attempt to straighten the elbows fully but at the same time, try to keep the pubic bone on the floor.
- If you are stiff, let the pelvis lift off the floor so that the body remains suspended. Do not let your hands slide further out to the front so that the pelvis can remain on the floor.
- Try to breathe into the stiffness, letting the spine drop down bit by bit as you remain there.
- Do not let your shoulders come up around your ears. Keep your neck long.
- Hold the position for one minute, breathing quietly as the spine relaxes.
- To come down, bend the elbows and lie the side of your face on the floor.
- Rest for 15 seconds and then repeat three times, trying to get the front of the pelvis as near as possible to the floor.

As a progression:

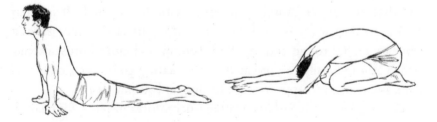

From the Cobra to the Pose of the Child

- When the hips fall through to the floor with little residual stiffness, you can move into a see-sawing action back and forth.
- Bend the hips by pulling your tummy in hard and pushing your bottom back towards your heels.
- With the top of your feet flat along the floor behind you, nestle your bottom down on your heels.
- Your hands will drag back along the floor a little way; leave them long and relaxed, elbows on the floor.
- Relax the side of your face or your forehead to the floor and remain there for five seconds.
- To rock forwards, slide your hands forward again and after lifting your bottom off your heels, let your hips fall through to the sagging position.
- Count five seconds in this position.
- Repeat four more times back and forth, counting all the time and trying to let the hips drop more each time.

The Sphinx

A milder version of the Cobra is the Sphinx exercise, where you lie prone, supported on your forearms, as if reading a book on your tummy. Although it does similar things to the Cobra, it is more effective at reversing a very stooped posture. Your back relaxes more in the Sphinx and thoraco-lumbar jamming can be specifically targeted. It is often painful to start with, as the humped middle part of the spine takes a while to drop through and sag. Your tummy and gluteal muscles may clench automatically to prevent the spine letting go but gradually this eases and the spine drops forward the longer you stay there.

Getting out of the Sphinx is often uncomfortable (even for healthy backs) and you may have to rock your hips from left to right, first gently and then with more gusto, before you can lift your pelvis back from the floor. It may be necessary to roll onto your back and ease your castness by rocking your knees to your chest. This becomes less necessary as your mobility improves. I often manually mobilise stiff thoraco-lumbar spines in the Sphinx position.

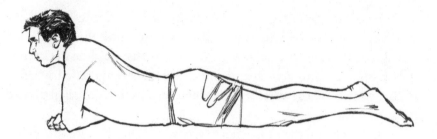

The correct way to do the Sphinx

- Lie prone on a carpeted floor with a pillow under your pelvis.
- Push up to support your upper body on both forearms, your elbows directly below your shoulders.
- Let your shoulders relax so your shoulder blades poke out and your spine is free to sag.
- Try to breathe into the stiffness and avoid tensing the spine as it moves through its jammed phase before it drops through to the floor.
- Don't hold onto your spine by tensing the tummy and gluteal muscles. Make them let go so the spine can drop down, one cog at a time.
- Hold the position for one minute and then gently begin to tip your pelvis from left to right in preparation for getting up.
- If you are not too stiff, you can push your bottom back to get up.
- If it hurts too much, roll over onto your back and rock your knees to your chest until the stiffness passes off.

Spinal intrinsics strengthening

This is the ultimate form of spinal strengthening. It is done face down, off the end of a table, with someone holding your feet as you lower your head to the floor and return to horizontal. It is the big brother to the strengthening part of toe touching in the standing position. But because there is a much longer leverage when the spine is suspended out in mid-air off the end of the table, the muscles work much harder to unfurl the spine. The proper action involves a humping movement of the lower back, followed by a wave-like undulation along the spine, with the head coming up last.

This is a very effective way of building up the strength of all the intrinsic muscles of the spine. Longissimus and iliocostalis are strengthened when they make the lumbar vertebrae slide backwards in a reverse gliding action and multifidus is strengthened when it pulls the tails of the vertebrae to slot them back from their tipped forward position.

As well as being the most effective way of strengthening the intrinsics, the horizontal exercise is also the most stable. By contrast, toe touches require more advanced coordination. If you have a very unstable segment you never start off from the vertical position, always the horizontal. You may be unlucky enough to get slippage of a vertebra while doing toe touches which will be painful, shake your confidence and leave the problem segment more irritable.

If you want brute strength for a spine which is weak you would go straight to the horizontal form of the intrinsics exercise, doing up to ten repetitions each session. But if you want to gradually build up the strength of a very weak, poorly controlled segment, you have to start off doing just one or two horizontal intrinsics each time. These strengthen the segment in its most stable position and then as control builds up, you progress to the toe touches from a standing position. By then the reaction time will have improved in the deep muscles and the defense mechanisms of the weak segment will be more advanced. When the general level of irritability of the problem segment is on the wane, the horizontal intrinsics provide the higher levels of strength and endurance.

Spinal intrinsics should *never* be done more often than once every week to ten days, otherwise your back will remain sore. The strenuous segment-shutting action of multifidus at the back of the interspaces can act as an irritant which keeps the back inflamed. Always follow each session with lengthy periods of knees rocking and spinal rolling but if your back takes more than a few days to settle, you should reduce the number of repetitions from ten to five. If the reaction is still too extreme, then you should extend the time period between sessions to two, possibly three weeks. You can only strengthen the muscles around a weak segment at the rate it can take it, otherwise your back will remain angrily on guard and you will sense no improvement at all.

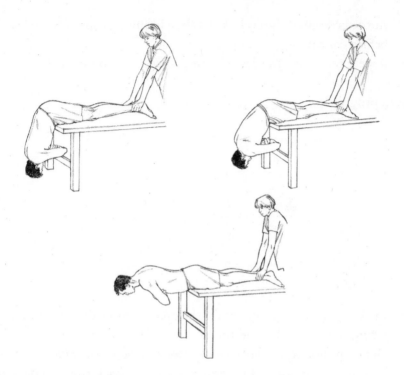

The correct way to do spinal intrinsics strengthening

- Lie face down on a firm surface such as a sturdy table.
- Place a pillow at the edge of the table and with somebody holding your feet, move up the table so your two hip bones are on the pillow.
- Allow your upper body to drop down towards the floor, putting a hand out to take your weight on the floor.
- Fold your arms across your chest and let your upper body hang down, making an angle of 90° at the hips.
- From this starting position, tighten your buttocks and pull your tummy in, rolling your pelvis back.
- Continuing this very powerful contraction, unfurl along the length of your spine one vertebra at a time, with your head coming up last. Do not attempt to hyperextend beyond the horizontal position.
- To return to the floor, duck the head down and do the movement in reverse.

- At the bottom of the cycle, hang for a moment, completely relaxed, before continuing.
- Repeat ten times, but less if your back reacts after each session.

As a progression:

- Increase the leverage and make the exercise more difficult by inter-lacing the fingers behind the neck with the elbows wide.
- Unfurl to horizontal *but do not hyperextend beyond that line.*
- You may do six of this version and as your back tires, go back to folding your arms on the chest for the other six.

Reference reading

Adams, M.A., *The mechanical environment of chondrocytes in articular cartilage Biorheology*, 2006; 43(3–4):537–45

Adams, M.A., Bogduk, N., Burton, K. and Dolan, P., *The Biomechanics of Back Pain*, Churchill Livingstone

Adams, M.A. and Dolan, P., 'Time-dependent changes in the lumbar spine's resistance to bending', *Clin Biomech* (Bristol, Avon) June 1996; 11(4): 194–2000

Adams, M.A., Dolan, P. and Hutton, W.C., 'The stages of disc degeneration as revealed by discograms', *Journal of Bone and Joint Surgery*, 1986; pp. 36–41

Adams, M.A., Dolan, P. and Hutton, W.C., 'Diurnal variations in the stresses on the lumbar spine', *Spine*, March 1987; 12(2):130–7

Adams, M.A., Dolan, P. and Hutton, W.C., 'The lumbar spine in backward bending', *Spine*, September 1988; 13(9):1019–26

Adams, M.A., Dolan, P., Hutton, W.C. and Porter, R.W., 'Diurnal changes in spinal mechanics and their clinical significance', *J Bone Joint Surg Br*, March 1990; 72(2):266–70

Adams, M.A., Freeman, B.J., Morrison, H.P., Nelson, I.W. and Dolan, P., 'Mechanical initiation of intervertebral disc degeneration', *Spine*, 1 July 2000; 25(13):1625–36

Adams, M.A. and Hutton, W.C., 'The effect of posture on the role of apophysial joints in resisting intervertebral compression forces', *J Bone Joint Surg Br*, August 1980; 62(3):358–62

Adams, M.A. and Hutton, W.C., 'The effect of posture on the fluid content of the lumbar intervertebral discs', *Spine*, September 1983; 8(6):665–71

Adams, M.A. and Hutton, W.C., 'The mechanical function of the lumbar apophyseal joints', *Spine*, April 1983; 8(3):327–30

Adams, M.A., McMillan, D.W., Green, T.P. and Dolan, P., 'Sustained loading generates stress concentrations in lumbar intervertebral discs', *Spine*, 15 February 1996; 21(4):434–8

Adams, M.A., McNally, D.S., Chinn, H. and Dolan, P., 'The Clinical Biomechanics Award Paper 1993 Posture and compressive strength of the lumbar spine', *Clinical Biomechanics* 1994; 9:5–14

Adams, M.A., McNally, D.S. and Dolan, P., '"Stress" distributions inside intervertebral discs: The effects of age and degeneration', *J Bone Joint Surg Br*, November 1996; 78(6):965–72

Adams, M.A., May, S., Freeman, B.J., Morrison, H.P. and Dolan, P., 'Effects of backward bending on lumbar intervertebral discs: Relevance to physical therapy treatments for low back pain', *Spine*, 15 February 2000; 25(4):431–7

Adams, M.A., Pollintine, P., Tobias, J.H., Wakley, G.K. and Dolan, P., 'Intervertebral disc degeneration can predispose to anterior vertebral fractures in the thoracolumbar spine', *J Bone Miner Res*, September 2006; 21(9):1409–16

Adams, M.A. and Roughley, P.J., 'What is intervertebral disc degeneration, and what causes it?', *Spine*, 15 August 2006; 31(18):2151–61

Allison, G.T. and Henry, S.M., 'Trunk muscle fatigue during a back extension task in standing', *Man Ther*, November 2001; 6(4):221–8

Anderson, D.G., Albert, T.J., Fraser, J.K., Risbud, M., Wuisman, P., Meisel, H.J., Tannoury, C., Shapiro, I, and Vaccaro, A.R., 'Cellular therapy for disc degeneration', *Spine*, 1 September 2005; 30(17 Suppl): S14–9

Arjmand, N. and Shirazi-Adl, A., 'Biomechanics of changes in lumbar posture in static lifting', *Spine*, 1 December 2005; 30(23):2637–48

Arjmand, N., Shirazi-Adl, A. and Bazrgari, B., 'Wrapping of trunk thoracic extensor muscles influences muscle forces and spinal loads in lifting tasks', *Clin Biomech* (Bristol, Avon), August 2006; 21(7):668–75

Atlas, S.J., Keller, R.B., Wu, Y.A., Deyo, R.A. and Singer, D.E., 'Long-term outcomes of surgical and nonsurgical management of lumbar spinal stenosis: 8 to 10 year results from the Maine lumbar spine study', *Spine*, 15 April 2005; 30(8):936–43

Axelsson, P. and Karlsson, B.S., 'Intervertebral mobility in the progressive degenerative process. A radiostereometric analysis', *Eur Spine*, October 2004; 13(6):567–72

Baranto, A., Ekstrom, L., Holm, S., Hellstrom, M., Hansson, H.A. and Sward, L., 'Vertebral fractures and separations of endplates after traumatic loading of adolescent porcine spines with experimentally-induced disc degeneration', *Clin Biomech* (Bristol, Avon), December 2005; 20(10):1046–54

Barker, P.J., Guggenheimer, K.T., Grkovic, I., Briggs, C.A., Jones, D.C., Thomas, C.D. and Hodges, P.W., 'Effects of tensioning the lumbar fasciae on segmental stiffness during flexion and extension: Young Investigator Award winner', *Spine*, 15 February 2006; 31(4):397–405

Basmajian, J.V., 'Acute back pain and spasm. A controlled multicentre trial of combined analgesic and anti-spasm agents', *Spine*, 1989; 14:438–9

Baylis, W.J., Rzonca, E.C., 'Functional and structural limb length discrepancies: evaluation and treatment', *Clin Podiatr Med Surg*, July 1988; 5(3):509–20

Beebe, F.A., Barkin, R.L. and Barkin, S., 'A clinical and pharmacologic review of skeletal muscle relaxants for musculoskeletal conditions', *Am J Ther*, March–April 2005; 12(2):151–71

Beneck, G.J., Kulig, K., Landel, R.F. and Powers, C.M., 'The relationship between lumbar segmental motion and pain response produced by a posterior-to-anterior force in persons with nonspecific low back pain', *Orthop Sports Phys Ther*, April 2005; 35(4):203–9

Benneker, L.M., Heini, P.F., Anderson, S.E., Alini, M. and Ito, K., 'Correlation of radiographic and MRI parameters to morphological and biochemical assessment of intervertebral disc degeneration', *Eur Spine J*, February 2005; 14(1):27–35

Bergmark, A., 'The stability of the lumbar spine: a study in mechanical engineering', *Acta Othopaedica Scandinavica*, 1989; 230(Suppl):20–24

Bibby, S.R., Fairbank, J.C., Urban, M.R. and Urban, J.P., 'Cell viability in scoliotic discs in relation to disc deformity and nutrient levels', *Spine*, 15 October 2002; 27(20):2220–8

Bogduk, N., 'Anatomy and biomechanics of psoas major', *Clinical Biomechanics* 1992; 7:109–119

Bogduk, N., *Clinical Anatomy of the Lumbar Spine and Sacrum*, Churchill Livingstone

Bogduk, N., 'The lumbar disc and low back pain', *Neurosurg Clin N Am*, October 1991; 2(4)791–800

Bogduk, N., 'Lumbar dorsal ramus syndrome', *Med J Aust*, 15 November 1980; 2(10):537–41

Bogduk, N., 'Management of chronic low back pain', *Med J* Aust, 19 January 2004; 180(2):79–83

Bogduk, N. and Jull, G., 'The theoretical pathology of acute locked back: A basis for manipulative therapy', *Manual Medicine*, 1985; 1:78–82

Boos, N., Rieder, R., Schade, V., Spratt, K.F., Semmer, N. and Aebi, M., Volvo Award in Clinical Sciences, 'The diagnostic accuracy of MRI, work perception and psychosocial factors in identifying symptomatic disc herniations', *Spine*, 1995; 20:2613–25

Borenstein, D.G. and Korn, S., 'Efficacy of a low-dose regimen of cyclobenzaprine hydrochloride in acute skeletal muscle spasm: results of two placebo-controlled trials', *Clin Ther*, April 2003; 25(4):1056–73

Boxberger, J.I., Sens, S., Yerramalli, C.S. and Elliott, D.M., 'Nucleus pulposus glycosaminoglycan content is correlated with axial mechanics in rat lumbar motion segments', *J Orthop Res*, 24 July 2006

Brinckmann, P., 'Injury of the annulus fibrosus and disc protrusions. An in vitro investigation on human lumbar discs', *Spine*, March 1986; 11(2):149–53

Brinckmann, P., Biggemann, M., Burton, K., Leirseth, G., Tillotson, M. and Frobin, W., 'Radiographic changes in the lumbar intervertebral discs and lumbar vertebrae with age', *Spine*, 1 January 2004; 29(1):108–9

Brinckmann, P., Biggemann, M. and Hilweg, D., 'Prediction of the compressive strength of the lumbar vertebrae', *Spine*, June 1989; 14(6):606–10

Brinckmann, P., Frobin, W., Hierholzer, E. and Horst, M., 'Deformation of the vertebral end-plate under axial loading of the spine', *Spine*, November–December 1983; 8(8):851–6

Brinckmann, P.and Grootenboer, H., 'Change of disc height, radial disc bulge, and intradiscal pressure from discectomy. An in vitro investigation on human lumbar discs', *Spine*, June 1991; 16(6):641–6

Burton, A.K., Battie, M.C., Gibbons, L., Videman, T. and Tillotson, K.M., 'Lumbar disc degeneration and sagittal flexibility', *J Spinal Disord*, October 1996; 9(5):418–24

Butler, D., Trafimow, J.H., Andersson, G.B., McNeill, T.W. and Huckman, M.S., 'Discs degenerate before facets', *Spine*, February 1990; 15(2):111–13

Chanchairujira, K., Chung, C.B., Kim, J.Y., Papakonstantinou, O., Lee, M.H., Clopton, P. and Resnick, D., 'Intervertebral disk calcification of the spine in an elderly population: radiographic prevalence, location, and distribution and correlation with spinal degeneration', *Radiology*, February 2004; 230(2):499–503

Childers, M.K., Borenstein, D., Brown, R.L., Gershon, S., Hale, M.E., Petri, M., Wan, G.J., Laudadio, C. and Harrison, D.D., 'Low-dose cyclobenzaprine versus combination therapy with ibuprofen for acute neck or back pain with muscle spasm: a randomized trial', *Curr Med Res Opin*, September 2005; 21(9):1485–93

Chiu, E.J., Newitt, D.C., Segal, M.R., Hu, S.S., Lotz, J.C. and Majumdar, S., 'Magnetic resonance imaging measurement of relaxation and water diffusion in the human lumbar intervertebral disc under compression in vitro', *Spine*, 1 October 2001; 26(19):E437–44

Cholewicki, J., Juluru, K., Radebold, A., Panjabi, M.M. and McGill, S.M., 'Lumbar spine stability can be augmented with an abdominal belt and/or increased intra-abdominal pressure', *Spine*, 15 August 2002; 27(16):1754

Cholewicki, J. and McGill, S.M., 'Mechanical stability of the in vivo lumbar spine: implications for injury and chronic low back pain', *Clinical Biomechanics* 1996; 11:1–15

Cholewicki, J. and Silfies, S.P., 'Clinical biomechanics of the lumbar spine', *Grieve's Modern Manual Therapy*, The Vertebral Column, Chap 7, 3rd edition, Elsevier, Churchill Livingstone

Cinotti, G., Della Rocca, C., Romeo, S., Vittur, F., Toffanin, R. and Trasimeni, G., 'Degenerative changes of porcine intervertebral disc induced by vertebral endplate injuries', *Spine*, 15 January 2005; 30(2):174–80.

Cooke, P.M. and Lutz, G.E., 'Internal disc disruption and axial back pain in the athlete', *Phys Med Rehabil Clin N Am*, November 2000; 11(4):837–65

Cordo, P.J., Gurfinkel, V.S., Smith, T.C., Hodges, P.W., Verschueren, S.M. and Brumagne, S., 'The sit-up: complex kinematics and muscle activity in voluntary axial movement', *J Electromyogr Kinesiol*, June 2003; 13(3):239–52

Court, C., Colliou, O.K., Chin, J.R., Liebenberg, E., Bradford, D.S. and Lotz, J.C., 'The effect of static in vivo bending on the murine intervertebral disc', *Spine*, July–August 2001; 1(4):239–45

Cresswell, A.G., Grundstrom, H. and Thorstensson, A., 'Observations on intra-abdominal pressure and patterns of abdominal intra-muscular activity in man', *Acta Physiol Scand*, April 1992; 144(4):409–18

Crock, H.V., AO, 'Internal disc disruption: A challenge to disc prolapse fifty years on', *Spine*, July–August 1986; 11(6):650–3

Cunningham, B.W., Gordon, J.D., Dmitriev, A.E., Hu, N. and McAfee, P.C., 'Biomechanical evaluation of total disc replacement arthroplasty: an in vitro human cadaveric model', *Spine*, 15 October 2003; 28(20):S110–7

Daggfeldt, K. and Thorstensson, A., 'The role of intra-abdominal pressure in spinal unloading', *J Biomech*, November–December 1997; 30(11–12):1149–55

Daggfeldt, K. and Thorstensson, A., 'The mechanics of back-extensor torque production about the lumbar spine', *J Biomech*, June 2003; 36(6):815–25

Derby, R., Lee, S.H., Kim, B.J., Chen, Y., Aprill, C. and Bogduk, N., 'Pressure-controlled lumbar discography in volunteers without low back symptoms', *Pain Med*, May–June 2005; 6(3):213–21

Dolan, P. and Adams, M.A., 'Recent advances in lumbar spinal mechanics and their significance for modelling', *Clin Biomech* (Bristol, Avon), 2001; 16(Suppl)1:S8–S16

Dolan, P., Kingma, I., van Dieen, J., Looze, M.P., Toussaint, H.M., Baten, C.T. and Adams, M.A., 'Dynamic forces acting on the lumbar spine during manual handling. Can they be estimated using electromyographic techniques alone?', *Spine*, 1 April 1999; 24(7):698–703

Donceel, P. and Du Bois, M., 'Fitness for work after surgery for lumbar disc herniation: a retrospective study', *Eur Spine*, 1998; J 7:29–35

Duncan, N.A., 'Cell deformation and micromechanical environment in the intervertebral disc', *J Bone Joint Surg Am*, April 2006; 88(Suppl)2:47–51

Dunlop, R.B., Adams, M.A. and Hutton, W.C., 'Disc space narrowing and the lumbar facet joints', *J Bone Joint Surg Br*, November 1984; 66(5):706–10

El-Rich, M., Shirazi-Adl, A. and Arjmand, N., 'Muscle activity, internal loads, and stability of the human spine in standing postures: combined model and in vivo studies', *Spine*, 1 December 2004; 29(23):2633–42

Essendrop, M., Andersen, T.B. and Schibye, B., 'Increase in spinal stability obtained at levels of intra-abdominal pressure and back muscle activity realistic to work situations', *Appl Ergon*, September 2002; 33(5):471–6

Farfan, H.F. and Sullivan, J.D., 'The relation of facet orientation to IVD failure', *Can J Surg*, 1967; 10:179–85

Farooq, N., Park, J.C., Pollintine, P., Annesley-Williams, D.J. and Dolan, P., 'Can vertebroplasty restore normal load-bearing to fractured vertebrae?', *Spine*, 1 August 2005; 30(15):1723–30

Ferguson, S.J., Ito, K. and Nolte, L.P., 'Fluid flow and convective transport of solutes within the intervertebral disc', *J Biomech*, February 2004; 37(2):213–21

Fisher, C.M., 'Pain states: a neurological commentary', *Clin Neurosurg* 1983; 31:32–53

Freemont, A.J., Peacock, T.E., Goupille, P., Hoyland, J.A., O'Brien, J. and Jayson, M-I.V., 'Nerve ingrowth into diseased intervertebral disc in chronic back pain', *Lancet*, 1997; 350:178–81

Friberg, O., 'Clinical symptoms and biomechanics of the lumbar spine and hip joint in leg length inequality', *Spine*, 1983; 8:643–51

Frobin, W., Brinckmann, P., Biggemann, M., Tillotson, M. and Burton, K., 'Precision measurement of disc height, vertebral height and sagittal plane displacement from lateral radiographic views of the lumbar spine', *Clin Biomech* (Bristol, Avon), 1997; 12(Suppl)1:S1–S63

Fryer, G., Morris, T. and Gibbons, P., 'Paraspinal muscles and intervertebral dysfunction: part one', *J Manipulative Physiol*, May 2004; 27(4):267–74

Fujiwara, A., Lim, T.H., An, H.S., Tanaka, N., Jeon, C.H., Andersson, G.B. and Haughton, V.M., 'The effect of disc degeneration and facet joint osteoarthritis on the segmental flexibility of the lumbar spine', *Spine*, 1 December 2000; 25(23):3036–44

Fujiwara, A., Tamai, K., An, H.S., Kurihashi, T., Lim, T.H., Yoshida, H. and Saotome, K., 'The relationship between disc degeneration, facet joint osteoarthritis, and stability of the degenerative lumbar spine', *J Spinal Disord*, October 2000; 13(5):444–50

Fujiwara, A., Tamai, K., Yamato, M., An, H.S., Yoshida, H., Saotome, K. and Kurihashi, A., 'The relationship between facet joint osteoarthritis and disc degeneration of the lumbar spine: an MRI study', *Eur Spine J*, 1999; 8(5):396–401

Ganey, T., Libera, J., Moos, V., Alasevic, O., Fritsch, K.G., Meisel, H.J. and Hutton, W.C., 'Disc chondrocyte transplantation in a canine model: A treatment for degenerated or damaged intervertebral disc', *Spine*, 1 December 2003; 28(23):2609–20

Gantenbein, B., Grunhagen, T., Lee, C.R., van Donkelaar, C.C., Alini, M. and Ito, K., 'An in vitro organ culturing system for intervertebral disc explants with vertebral endplates: a feasibility study with ovine caudal discs', *Spine*, 1 November 2006; 31(23):2665–73

Gardner-Morse, M.G. and Stokes, I.A.F., 'The effect of abdominal muscle co-activation on lumbar spine stability', *Spine*, 1 January 1998; 23(1):86–91; discussion 91–2

Gardner-Morse, M.G. and Stokes, I.A.F, 'Structural behavior of human lumbar spinal motion segments', *J Biomech*, February 2004; 37(2):205–12

Gibson, J.N., Grant, I.C. and Waddell, G., 'The Cochrane review of surgery for lumbar disc prolapse and degenerative lumbar spondylosis', *Spine*, 1999; 24:1820–32

Goto, K., Tajima, N., Chosa, E., Totoribe, K., Kubo, S., Kuroki, H. and Arai, T., 'Effects of lumbar spinal fusion on the other lumbar intervertebral levels (three-dimensional finite element analysis)', *J Orthop Sci*, 2003; 8(4):577–84

Gower, W.E. and Pedrini, V., 'Age related variation in protein polysaccharides from human nucleus pulposis, annulus fibrosis and costal cartilage', *Journal of Bone and Joint Surg Am*, 1969; 51A:1154–62

Green, T.P., Allvey, J.C. and Adams, M.A., 'Spondylolysis. Bending of the inferior articular processes of lumbar vertebrae during simulated spinal movements', *Spine*, 1 December 1994; 19(23):2683–91

Grobler, L.J., Novotny, J.E., Wilder, D.G., Frymoyer, J.W. and Pope, M.H., 'L4-5 isthmic spondylolisthesis. A biomechanical analysis comparing stability in L4-5 and L5-S1 isthmic spondylolisthesis', *Spine*, 15 January 1994; 19(2):222–7

Grundy, P.F. and Roberts, C.J., 'Does unequal leg length cause back pain? A case-control study', *Lancet*; iv:256–8

Grunhagen, T., Wilde, G., Soukane, D.M., Shirazi-AdlS, A. and Urban, J.P., 'Nutrient supply and intervertebral disc metabolism', *J Bone Joint Surg Am*, April 2006; 88(Suppl)2:30–5

Guehring, T., Omlor, G.W., Lorenz, H., Engelleiter, K., Richter, W., Carstens, C. and Kroeber, M., 'Disc distraction shows evidence of regenerative potential in degenerated intervertebral discs as evaluated by protein expression, magnetic resonance imaging, and messenger ribonucleic acid expression analysis', *Spine*, 1 July 2006; 31(15):1658–65

Guehring, T., Unglaub, F., Lorenz, H., Omlor, G., Wilke, H.J. and Kroeber, M.W., 'Intradiscal pressure measurements in normal discs, compressed discs and compressed discs treated with axial posterior disc distraction: an experimental study on the rabbit lumbar spine model', *Eur Spine J*, May 2006; 15(5):597–604

Gurney, B., 'Leg length discrepancy', *Gait Posture*, April 2002; 15(2):195–206

Haefeli, M., Kalberer, F., Saegesser, D., Nerlich, A.G., Boos, N. and Paesold, G., 'The course of macroscopic degeneration in the human lumbar intervertebral disc', *Spine*, 15 June 2006; 31(14):1522–31

Hake, A., Frobin, W., Brinckmann, P. and Biggemann, M., 'Sagittal plane rotational and translational motion of lumbar segments with decreascd intervertebral disc height', *Rofo*, August 2002; 174(8):996–1002

Handa, T., Ishihara, H., Ohshima, H., Osada, R., Tsuji, H. and Obata, K., 'Effects of hydrostatic pressure on matrix synthesis and matrix metal-loproteinase production in the human lumbar intervertebral disc', *Spine*, 15 May 1997; 22(10):1085–91

Harrison, D.E., Caillet, R., Harrison, D.D., Janik, T.J. and Holland, B., 'Changes in lumbar sagittal configuration with a new method of extension traction: Non randomized clinical control trial', *Arch Phys Med Rehabil*, 2002; 38:1585–91

Hedman, T.P. and Fernie, G.R., 'Mechanical response of the lumbar spine to seated postural loads', *Spine*, 1 April 1997; (7):734–4522

Hides, J.A., Richardson, C.A. and Jull, G.A., 'Multifidus muscle recovery is not automatic after resolution of acute, first episode low back pain', *Spine*, 1 December 1996; 21(23):2763–9

Hides, J., Wilson, S., Stanton, W., McMahon, S., Keto, H., McMahon, K., Bryant, M. and Richardson, C., 'An MRI investigation into the function of the transversus abdominis muscle during "drawing-in" of the abdominal wall', *Spine*, 15 March 2006; 31(6)

Hirayama, J., Yamagata, M., Ogata, S., Shimizu, K., Ikeda, Y. and Taka-hashi, K., 'Relationship between low-back pain, muscle spasm and pressure pain thresholds in patients with lumbar disc herniation', *Eur Spine J*, February 2006; 15(1):41–7

Hodges, P.W., 'Motor control of the trunk', *Grieve's Modern manual Therapy*, Chap 10, 3rd edition, Elsevier, Churchill Livingstone

Hodges, P.W., 'The role of the motor system in spinal pain: implications for rehabilitation of the athlete following lower back pain', *J Sci Med Sport*, September 2000; 3(3):243–53

Hodges, P.W., Kaigle Holm, A., Holm, S., Ekstrom, L., Cresswell, A., Hansson, T. and Thorstensson, A., 'Intervertebral stiffness of the spine is increased by evoked contraction of transversus abdominus and the diaphragm: in vivo porcine studies', *Spine*, 2003; 29(23):2594–601

Hodges, P.W. and Richardson, C.A., 'Inefficient muscular stabilization of the lumbar spine associated with low back pain. A motor control evaluation of transversus abdominis', *Spine*, 15 November 1996; 21(22):2640–50

Holm, S., Holm, A.K., Ekstrom, L., Karladani, A. and Hansson, T., 'Exper-imental disc degeneration due to endplate injury', *J Spinal Disord Tech*, February 2004; 17(1):64–71

Holmes, A.D., Hukins, D.W. and Freemont, A.J., 'End-plate displacement during compression of lumbar vertebra-disc-vertebra segments and the mechanism of failure', *Spine*, January 1993; 18(1):128–35

Horner, H.A. and Urban, J.P., '2001 Volvo Award Winner in Basic Science Studies: Effect of nutrient supply on the viability of cells from the nucleus pulposus on the intervertebral disc', *Spine*, 1 December 2001; 26(23):2543–9

Hu, J.C. and Athanasiou, K.A., 'The effects of intermittent hydrostatic pressure on self-assembled articular cartilage constructs', *Tissue Eng*, 1 May 2006

Hutton, W.C., Elmer, W.A., Bryce, L.M., Kozlowska, E.E., Boden, S.D. and Kozlowski, M., 'Do the intervertebral disc cells respond to different levels of hydrostatic pressure?', *Clin Biomech* (Bristol, Avon), November 2001; 16(9):728–34

Hutton, W.C., Ganey, T.M., Elmer, W.A., Kozlowska, E., Ugbo, J.L., Doh, E.S. and Whitesides, T.E. Jr., 'Does long-term compressive loading on the intervertebral disc cause degeneration?', *Spine*, 1 December 2000; 25(23):2993–3004

Hutton, W.C., Malko, J.A. and Fajman, W.A., 'Lumbar disc volume measured by MRI: effects of bed rest, horizontal exercise, and vertical loading', *Aviat Space Environ Med*, January 2003; 74(1):73–8

Hutton, W.C., Murakami, H., Li, J., Elmer, W.A., Yoon, S.T., Minamide, A., Akamura, T. and Tomita, K., 'The effect of blocking a nutritional pathway to the intervertebral disc in the dog model', *J Spinal Disord Tech*, February 2004; 17(1):53–63

Hutton, W.C., Yoon, S.T., Elmer, W.A., Li, J., Murakami, H., Minamide, A. and Akamuru, T., 'Effect of tail suspension (or simulated weightlessness) on the lumbar intervertebral disc: study of proteoglycans and collagen', *Spine*, 15 June 2002; 27(12):1286–90

Iatridis, J.C., Maclean, J.J., Roughley, P.J. and Alini, M., 'Effects of Mechanical loading on IVD metabolism in vivo', *Journal of Bone and Joint Surgery Am*, April 2006; 88(Suppl)2:41–6

Iatridis, J.C., Mente, P.L., Stokes, I.A., Aronsson, D.D. and Alini, M., 'Compression-induced changes in intervertebral disc properties in a rat tail model', *Spine*, 15 May 1999; 24(10):996–1002

Indahl, A., Kaigle, A.M., Reikeras, O. and Holm, S.H., 'Interaction between the porcine lumbar intervertebral disc, zygapophysial joints and paraspinal muscles', *Spine*, 15 December 1997; 22(24):2834–40

Inoue, G., Ohtori, S., Aoki, Y., Ozawa, T., Doya, H., Saito, T., Ito, T., Akazawa, T., Moriya, H. and Takahashi, K., 'Exposure of the nucleus pulposis to the outside of the anulus fibrosis induces nerve injury and regeneration of the afferent fibres innervating the lumbar intervertebral discs in rats', *Spine*, 1 June 2006; 31(13):1433–8

Ishihara, H., McNally, D.S., Urban, J.P. and Hall, A.C., 'Effects of hydrostatic pressure on matrix synthesis in different regions of the intervertebral disc', *J Appl Physiol*, March 1996; 80(3):839–46

Ishihara, H. and Urban, J.P., 'Effects of low oxygen concentrations and metabolic inhibitors on proteoglycan and protein synthesis rates in the intervertebral disc', *Journal of Orthopaedic Research*, 1999; 17:829–35

Iwamoto, J., Abe, H., Tsukimura, Y. and Wakano, K., 'Relationship between radiological abnormalities of the lumbar spine and incidence of LBP in high school rugby players: a prospective study', *Scan J Med Sci Sports*, June 2005; 15(3):163–8

Jensen, M.C., Brant-Zawadzki, M.N., Obuchowski, N., Modic, M.T., Malkasian, D. and Ross, J.S., 'Magnetic resonance imaging of the lumbar spine in people without back pain', *New England Journal of Medicine*, July 1994; 331(2):69–73

Johannessen, W., Auerbach, J.D., Wheaton, A.J., Kurji, A., Borthakur, A., Reddy, R. and Elliott, D.M., 'Assessment of human disc degeneration and proteoglycan content using T1rho-weighted magnetic resonance imaging', *Spine*, 15 May 2006; 31(11):1253–7

Kaigle, A.M., Holm, S.H. and Hansson, T.H., 'Experimental instability in the lumbar spine', *Spine*, 15 February 1995; 20(4):421–30

Kaigle, A.M., Holm, S.H. and Hansson, T.H., '1997 Volvo Award winner in biomechanical studies. Kinematic behaviour of porcine lumbar spine: a chronic lesion model', *Spine*, 15 December 1997; 22(24):2796–806

Katz, M.M., Hargens, A.R. and Garfin, S.R., 'Intervertebral disc nutrition. Diffusion versus convection', *Clin Orthop Relat Res*, September 1986; (210):243–5

Kazarian, L.E., 'Creep characteristics of the human vertebral column', *Orthop Clin North Am*, 1975; 6:3–8

Keifer, A., Shirazi-Adl, A. and Parnianpour, M., 'Synergy of the human spine in neutral positions', *European Spine Journal*, 1998; 7:471–9

Keller, T.S., Colloca, C.J., Harrison, D.E., Harrison, D.D. and Janik, T.J., 'Influence of spine morphology on intervertebral disc loads and stresses in asymptomatic adults: implications for the ideal spine', *Spine J*, May–June 2005; 5(3):297–309

Keller, T.S. and Nathan, M., 'Height change caused by creep in intervertebral discs: a sagittal plane model', *J Spinal Disord*, August 1999; 12(4):313–24

Khan, A.M., Girardi, F., 'Spinal lumbar synovial cysts. Diagnosis and management challenge', *Eur Spine J*, August 2006; 15(8):1176–82

Kirkaldy-Willis, W.H. and Farfan, H.F., 'Instability of the lumbar spine', *Clin Orthop Relat Res*, May 1982; (165):110–23

Kniesel, B., 'Lumbar discogenic pain', *Z Orthop Ihre Grenzgeb*, November–December 2004; 142(6):709–15

Koumantakis, G.A., Watson, P.J. and Oldham, J.A., 'Trunk muscle stabilization training plus general exercise versus general exercise only: randomized controlled trial of patients with recurrent low back pain', *Phys Ther*, March 2005; 85(3):209–25

Kraemer, J., Colditz, D. and Gowin, R., 'Water and electrolyte content of human intervertebral discs under variable load', *Spine*, 1985; 10:69–71

Kroeber, M., Unglaub, F., Guehring, T., Nerlich, A., Hadi, T., Lotz, J. and Carstens, C., 'Effects of controlled dynamic disc distraction on degenerated intervertebral discs: an in vivo study on the rabbit lumbar spine model', *Spine*, 15 January 2005; 30(2):181–7

Kroeber, M.W., Unglaub, F., Wang, H., Schmid, C., Thomsen, M., Nerlich, A. and Richter, W., 'New in vivo animal model to create intervertebral disc degeneration and to investigate the effects of therapeutic strategies to stimulate disc regeneration', *Spine*, 1 December 2002; 27(23):2684–90

Kuroki, H., Goel, V.K., Holekamp, S.A., Ebraheim, N.A., Kubo, S. and Tajima, N., 'Contributions of flexion-extension cyclic loads to the lumbar spinal segment stability following different discectomy procedures', *Spine*, February 2004; 1(3):29

Kuslich, S.D., Ulstrom, C.L. and Michael, C.J., 'The tissue origin of low back pain and sciatica: a report of pain response to tissue stimulation during operations on the lumbar spine using local anesthesia', *Orthop Clin North Am*, April 1991; 22(2):181–7

Lorimer Moseley, G., Hodges, P. and Gandevia, S., 'Deep and superficial fibres of the lumbar multifidus muscle are differentially active during voluntary arm movements', *Spine*, 2002; 27(2):E29–E36

Lotz, J.C., 'Animal models of intervertebral disc degeneration: lessons learned', *Spine*, 1 December 2004; 29(23):2742–50

Lotz, J.C., Colliou, O.K., Chin, J.R., Duncan, N.A. and Liebenberg, E., 'Compression induced degeneration of the IVD: an in vivo mouse model and finite element study', *Spine*, 1 December 1998; 23(23):2495–506

Lotz, J.C., Kim, A.J., 'Disc regeneration: why, when, and how', *Neurosurg Clin N Am*, October 2005; 16(4):657–63, vii

Luk, K.D., Chow, D.H. and Holmes, A., 'Vertical instability in spondylolisthesis: a traction radiographic assessment technique and the principle of management', *Spine*, 15 April 2003; 28(8):819–27

Lundon, K. and Bolton, K., 'Structure and function of the lumbar intervertebral disc in health, aging and pathologic conditions', *J Orthop Sports Phys Ther*, June 2001; 31(6):291–303

Luoma, K., Vehmas, T., Raininko, R., Luukkonen, R. and Riihimaki, H., 'Lumbosacral transitional vertebra: relation to disc degeneration and low back pain', *Spine*, 15 January 2004; 29(2):200–5

MacDonald, D.A., Moseley, G.L. and Hodges, P.W., 'The lumbar multifidus: does the evidence support clinical beliefs?', *Man Ther*, November 2006; 11(4):254–63

Magnusson, M.L., Aleksiev, A.R., Spratt, K.F., Lakes, R.S. and Pope, M.H., 'Hyperextension and spine height changes', *Spine*, 1996; 21:2670–5

Malko, J.A., Hutton, W.C. and Fajman, W.A., 'An in vivo magnetic resonance imaging study of changes in the volume (and fluid content) of the lumbar intervertebral discs during a simulated diurnal load cycle', *Spine*, 15 May 1999; 24(10):1015–22

Malter, A.D., McNeney, B., Loeser, J.D. and Deyo, R.A., '5-year reoperation rates after different types of lumbar spine surgery', *Spine*, 1 April 1998; 23(7):814–20

Mannion, A.F., Adams, M.A. and Dolan, P., 'Sudden and unexpected loading generates high forces on the lumbar spine', *Spine*, 1 April 2000; 25(7):842–52

Markolf, K.L. and Morris, J.M., 'Deformation of the thoraco-lumbar intervertebral joints in response to external loads', *Journal of Bone and Joint Surgery*, 1972; 54A:511–33

McFadden, K.D. and Taylor, J.R., 'Axial rotation and the lumbar spine and gaping of the zygapophyseal joints', *Spine*, April 1990; 15(4):295–9

McMillan, D.W., Garbutt, G. and Adams, M.A., 'Effect of sustained loading on the water content of intervertebral discs: implications for disc metabolism', *Ann Rheum Dis*, December 1996; 55(12):880–7

McNab, I., 'Cervical Pain', *The mechanism of spondylogenic pain*, Pergamon Press, 1972, p. 196

McNally, D.S., Shackleford, I.M., Goodship, A.E. and Mulholland, R.C., 'In vivo stress measurement can predict pain on discography', *Spine*, 15 November 1996; 21(22):2580–7

Meisel, H.J., Ganey, T., Hutton, W.C., Libera, J., Minkus, Y. and Alasevic, O., 'Clinical experience in cell-based therapeutics: intervention and outcome', *Eur Spine J*, 19 July 2006

Mercer, S., 'Kinematics of the spine', *Grieve's Modern Manual Therapy*, Chap 4, 3rd edition, Elsevier, Churchill Livingstone

Milanese, S., 'Clinic biomechanics of lifting', *Grieve's Modern Manual Therapy*, Chap 8, 3rd edition, Elsevier, Churchill Livingstone

Miyamoto, H., Doita, M., Nishida, K., Yamamoto, T., Sumi, M. and Kurosaka, M., 'Effects of cyclic mechanical stress on the production of inflammatory agents by nucleus pulposus and anulus fibrosus derived cells in vitro', *Spine*, 1 January 2006; 31(1):10

Mochida, J., 'New strategies for disc repair: Novel preclinical trials', *J Orthop Sci*, 2005; 10(1):112–18

Mooney, V. and Robertson, J., 'The facet syndrome', *Clin Orthop*, 1976; 115:149–56

Moore, R.J., 'The vertebral endplate: Disc degeneration, disc regeneration', *Eur Spine J*, 1 July 2006

Moseley, G.L. and Hodges, P.W., 'Are the changes in postural control associated with low back pain caused by pain interference?', *Clin J Pain*, July–August 2005; 21(4):323–9

Moseley, G.L., Hodges, P.W. and Gandevia, S.C., 'Deep and superficial fibers of lumbar multifidus muscle are differentially active during voluntary arm movements', *Spine*, 15 January 2002; 27(2):E29–36

Moseley, G.L., Hodges, P.W. and Gandevia, S.C., 'External perturbation of the trunk in standing humans differentially activates components of the medial back muscles', *J Physiol*, 1 March 2003; 547(Pt 2):581–7; Epub, 20 December 2002

Nachemson, A.L., 'Disc pressure measurements', *Spine*, 1981; 6:93–7

Nachemson, A.L, 'In vitro diffusion of dye through the end-plates and the annulus fibrosis in the human IVD', *Acta Orthopaedica Scandinavica*, 1970; 41

Nachemson, A.L., 'The lumbar spine: an orthopaedic challenge', *Spine*, 1976; 1:59–71

Nachemson, A.L, 'Towards a better understanding of low-back pain: A review of the mechanics of the lumbar disc', *Rheumatol Rehabil*, August 1975; 14(3):129–43

Natarajan, R.N., Williams, J.R. and Andersson, G.B., 'Modeling changes in intervertebral disc mechanics with degeneration', *J Bone Joint Surg Am*, April 2006; 88(Suppl)2:36–40

Ng, J.K., Parnianpour, M., Richardson, C.A. and Kippers, V., 'Effect of fatigue on torque output and electroyographic measures of trunk muscles during isometric axial rotation', *Arch Phys Med Rehabil*, March 2003; 84(3):374–81

Noren, R., Trafimow, J., Andersson, G.B. and Huckman, M.S., 'The role of facet joint tropism and facet angle in disc degeneration', *Spine*, May 1991; 16(5):530–2

O'Hara, B.P., Urban, J.P. and Maroudas, A., 'Influence of cyclic loading on the nutrition of articular cartilage', *Ann Rheum Dis*, July 1990; 49(7):536–9

Ohshima, H. and Urban, J.P., 'The effect of lactate and pH on proteoglycans and protein synthesis rates in the intervertebral disc', *Spine*, September 1992; 17(9):1079–82

Olmarker, K., Blomquist, J., Stromberg, J., Nannmark, U., Thomsen, P. and Rydevik, B., 'Inflammatogenic properties of nucleus pulposis', *Spine*, 1995; 20:665–9

O'Sullivan, P.B., 'Clinical instability of the lumbar spine: its pathological basis, diagnosis and conservative management', *Grieve's Modern Manual Therapy*, The Vertebral Column, Chap 22, 3rd edition, Elsevier, Churchill Livingstone

O'Sullivan, P.B, 'Diagnosis and classification of chronic low back pain disorders: maladaptive movement and motor control impairments as underlying mechanism', *Man Ther*, November 2005; 10(4):242–55

O'Sullivan, P.B., 'Lumbar segmental 'instability': clinical presentation and specific stabilizing exercise management', *Man Ther*, February 2000; 5(1):2–12

Panjabi, M.M., 'A hypothesis of chronic back pain: ligament subfailure injuries lead to muscle control dysfunction', *Eur Spine J*, May 2006; 15(5):668–76

Panjabi, M.M., 'Clinical spinal instability and low back pain', *J Electromyogr Kinesiol*, August 2003; 13(4):371–9

Panjabi, M.M., 'The stabilizing system of the spine. Part I. Function, dysfunction, adaptation, and enhancement', *J Spinal Disord*, December 1992; 5(4):383–9

Panjabi, M.M., 'The stabilizing system of the spine. Part II. Neutral zone and instability hypothesis', *J Spinal Disord*, December 1992; 5(4):390–6

Paris, S.V., 'Physical signs of instability', *Spine*, April 1985; 10(3):277–9

Parkkinen, J.J., Ikonen, J., Lammi, M.J., Laakkonen, J., Tammi, M. and Helminen, H.J., 'Effects of cyclic hydrostatic pressure on proteoglycan synthesis in cultured chondrocytes and articular cartilage explants', *Arch Biochem Biophys*, January 1993; 300(1):458–65

Peng, B., Hao, J., Hou, S., Wu, W., Jiang, D., Fu, X. and Yang, Y., 'Possible pathogenesis of painful intervertebral disc degeneration', *Spine*, 1 March 2006; 31(5):560–6

Pflaster, D.S., Krag, M.H., Johnson, C.C., Haugh, L.D. and Pope, M.H., 'Effect of test environment on intervertebral disc hydration', *Spine*, 15 January 1997; 22(2):133–9

Pollintine, P., Dolan, P., Tobias, J.H. and Adams, M.A., 'Intervertebral disc degeneration can lead to 'stress-shielding' of anterior vertebral body: a cause of osteoporotic vertebral fracture?', *Spine*, 1 April 2004; 29(7):774–82

Pollintine, P., Przybyla, A.S., Dolan, P. and Adams, M.A., 'Neural arch load-bearing in old and degenerated spines', *J Biomech*, February 2004; 37(2):197–204

Powers, C.M., Kulig, K. and Bergman, G., 'Segmental mobility of the lumbar spine during a posterior to anterior mobilization: assessment using dynamic MRI', *Clin Biomech* (Bristol, Avon), January 2003; 18(1):80–3

Przybyla, A., Pollintine, P., Bedzinski, R. and Adams, M.A., 'Outer annulus tears have less effect than endplate fracture on stress distributions inside intervertebral discs: Relevance to disc degeneration', *Clin Biomech* (Bristol, Avon), December 2006; 21(10):1013–19

Quack, C., Schenk, P., Laeubli, T., Spillmann, S., Hodler, J., Michel, B.A. and Klipstein, A., 'Do MRI findings correlate with mobility tests? An explorative analysis of the test validity with regard to structure', *Eur Spine J*, 2 December 2006

Quint, U., Wilke, H.J., Shirazi-Adl, A., Parnianpour, M., Loer, F. and Claes, L.E., 'Importance of intersegmental trunk muscles for the stability of the lumbar spine. A biomechanical study in vitro', *Spine*, 15 September 1998; 23(18):1937–45

Rao, R.D., Wang, M., Singhal, P., McGrady, L.M. and Rao, S., 'Intradiscal pressure and kinematic behavior of lumbar spine after bilateral laminotomy and laminectomy', *Spine*, September–October 2002; 2(5): 320–6

Richardson, C.A. and Hides, J.A., 'The rationale of a motor control programme for the treatment of spinal muscle dysfunction', *Grieve's Modern Manual Therapy*, Chap 31, 3rd edition, Elsevier, Churchill Livingstone

Roberts, S., Evans, H., Trivedi, J. and Menage, J., 'Histology and pathology of the human intervertebral disc', *J Bone Joint Surg Am*, April 2006; 88(Suppl)2:10–4

Rohlmannt, A., Claes, L.E., Bergmannt, G., Graichen, F., Neef, P. and Wilke, H.J., 'Comparison of intradiscal pressures and spinal fixator loads for different body positions and exercises, *Ergonomics*, June 2001; 20:44(8):781–94

Saraste, H., 'Long term clinical and radiological followup of spondylolysis and spondylolisthesis', *Journal of Paediatric Orthopaedics*, 1987; 7:631

Sato, K., Kikuchi, S. and Yonezawa, T., 'In vivo intradiscal pressure measurement in healthy individuals and in patients with ongoing back problems', *Spine*, 1 December 1999; 24(23):2468–74

Schnake, K.J., Putzier, M., Haas, N.P. and Kandziora, F., 'Mechanical concepts for disc regeneration', *Eur Spine J*, 12 July 2006

Schneider, G., Pearcy, M.J. and Bogduk, N., 'Abnormal motion in spondylolytic spondylolisthesis', *Spine*, 15 May 2005; 30(10):1159–64

Schwarzer, A.C., Aprill, C.N., Derby, R., Fortin, J., Kine, G. and Bogduk, N., 'The prevalence and clinical features of internal disc disruption in patients with chronic low back pain', *Spine*, 1 September 1995; 20(17):1878–83

Selard, E., Shirazi-Adl, A. and Urban, J.P., 'Finite element study of nutrient diffusion in the human intervertebral disc', *Spine*, 1 September 2003; 28(17):1945–53

Setton, L.A. and Chen, J., 'Mechanobiology of the intervertebral disc and relevance to disc degeneration', *Bone Joint Surg Am*, April 2006; 88(Suppl)2:52–7

Shirazi-Adl, A., 'Finite-element simulation of changes in the fluid content of human lumbar discs. Mechanical and clinical implications', *Spine*, February 1992; 17(2):206–12

Shirazi-Adl, A., Sadouk, S., Parnianpour, M., Pop, D. and El-Rich, M., 'Muscle force evaluation and the role of posture in human lumbar spine under compression', *Eur Spine J*, December 2002; 11(6):519–26

Singer, K.P., 'The spine and the effect of ageing', *Grieve's Modern Manual Therapy*, The Vertebral Column, Chap 14, 3rd edition, Elsevier, Churchill Livingstone

Sivan, S., Neidlinger-Wilke, C., Wurtz, K., Maroudas, A. and Urban, J.P., 'Diurnal fluid expression and activity of intervertebral disc cells', *Biorheology*, 2006; 43(34):283–91

Snijders, C.J., Hermans, P.F., Niesing, R., Spoor, C.W. and Stoeckart, R., 'The influence of slouching and lumbar support on iliolumbar ligaments, intervertebral discs and sacroiliac joints', *Clin Biomech*, May 2004; 19(4):323–9

Sobajima, S., Kompel, J.F., Kim, J.S., Wallach, C.H., Robertson, D.D., Vogt, M.T., Kang, J.D. and Gilberston, L.G., 'A slowly progressive and reproducible animal model of intervertebral disc degeneration characterized by MRI, X-ray, and histology', *Spine*, 1 January 2005; 30(1):15–24

Stokes, I.A. and Iatridis, J.C., 'Mechanical conditions that accelerate intervertebral disc degeneration: overload versus immobilization', *Spine*, 1 December 2004; 29(23):2724–32

Takahashi, I., Kikuchi, S., Sato, K. and Sato, N., 'Mechanical load of the lumbar spine during forward bending motion of the trunk-a biomechanical study', *Spine*, 1 January 2006; 31(1):18–23

Tanaka, N., An, H.S., Lim, T.H., Fujiwara, A., Jeon, C.H. and Haughton, V.M., 'The relationship between disc degeneration and flexibility of the lumbar spine', *Spine J*, January–February 2001; 1(1):47–56

Ten Brinke, A., van der Aa, H.E., van der Palen, J. and Oosterveld, F., 'Is leg length discrepancy associated with the side of radiating pain in patients with lumbar herniated disc?', *Spine*, 1 April 1999; 24(7):684–6

Thornton, W.E., Moore, T.P. and Pool, S.L., 'Fluid shifts in weightlessness', *Aviation, Space and Environmental Medicine*, 1987; 58:A86–A90

Tofferi, J.K., Jackson, J.L. and O'Malley, P.G., 'Treatment of fibromyalgia with cyclobenzaprine: A meta-analysis', *Arthritis Rheum*, 15 February 2004; 51(1):9–13

Toth, P.P. and Urtis, J., 'Commonly used muscle relaxant therapies for acute low back pain: a review of carisoprodol, cyclobenzaprine hydrochloride, and metaxalone', *Clin Ther*, September 2004; 26(9):1355–67

Tsantrizos, A., Ito, K., Aebi, M. and Steffen, T., 'Internal strains in healthy and degenerated lumbar intervertebral discs', *Spine*, 1 October 2005; 30(19):2129–37

Twomey, L.T., 'Sustained lumbar traction. An experimental study of long spine segments', *Spine*, March 1985; 10(2):146–9

Twomey, L.T. and Taylor, J.R., 'Age changes in lumbar vertebrae and intervertebral discs', *Clin Orthop Relat Res*, November 1987; (224):97–104

Twomey, L.T. and Taylor, J.R., *Physical Therapy of the Low Back*, 3rd edition, 2000, Churchill Livingstone, New York

Tyrrell, A.R., Reilly, T. and Troup, J.D., 'Circadian variation in stature and the effects on spinal loading', *Spine*, March 1985; 10(2):161–4

Unglaub, F., Guehring, T., Lorenz, H., Carstens, C. and Kroeber, M.W., 'Effects of unisegmental disc compression on adjacent segments: an in vivo animal model', *Eur Spine J*, December 2005; 14(10):949–55

Unglaub, F., Guehring, T., Omlor, G., Lorenz, H., Carstens, C. and Kroeber, M.W., 'Controlled distraction as a therapeutic option in moderate degeneration of the intervertebral disc—an in vivo study in the rabbit-spine model', *Z Orthop Ihre Grenzgeb*, January–February 2006; 144(1):68–73

Urban, J.P., Holm, S., Maroudas, A. and Nachemson, A., 'Nutrition of the intervertebral disc: effect of fluid flow on solute transport', *Clin Orthop Relat Res*, October 1982; (170):296–302

Urban, J.P. and Maroudas, A., 'Swelling of the intervertebral disc in vitro', *Connective Tissue Research*, 1981; 9:1–10

Urban, J.P. and Roberts, S., 'Chemistry of the IVD in relation to functional requirements', *Grieve's Modern manual Therapy*, 'The Vertebral Column', Chap 5, 3rd edition, Elsevier, Churchill Livingstone

Urban, M.R., Fairbank, J.C., Etherington, P.J., Loh, L., Winlove, C.P. and Urban, J.P., 'Electrochemical measurement of transport into scoliotic intervertebral discs in vivo using nitrous oxide as a tracer' *Spine*, 15 April 2001; 26(8):984–90

Urquhart, D.M. and Hodges, P.W., 'Differential activity of regions of transversus abdominis during trunk rotation', *Eur Spine J*, May 2005; 14(4):393–400

Urquhart, D.M., Hodges, P.W., Allen, T.J. and Story, I.H., 'Abdominal muscle recruitment during a range of voluntary exercises', *Man Ther*, May 2005; 10(2):144–53

Van Tulder, M.W., Touray, T., Furlan, A.D., Solway, S., Bouter and L.M., 'Muscle relaxants for nonspecific low back pain: a systematic review within the framework of the cochrane collaboration', *Spine*, 1 November 2004; 29(21):2474

Vernon-Roberts, B. and Pirie, C.J., 'Degenerative changes in the intervertebral discs of the lumbar spine and their sequelae', *Rheumatol Rehabil*, February 1977; 16(1):13–21

Waldman, S.D., Spiteri, C.G., Grynpas, M.D., Pilliar, R.M. and Kandel, R.A., 'Long-term intermittent compressive stimulation improves the composition and mechanical properties of tissue-engineered cartilage', *Tissue Eng*, September–October 2004; 10(9–10):1323–31

Walsh, A.J. and Lotz, J.C., 'Biological response of the intervertebral disc to dynamic loading', *J Biomech*, March 2004; 37(3):329–37

Wardell, G., Nachemson, A. and Phillips, R.B., *The Back Pain Revolution*, 1998, Churchill, Livingstone

Wilke, H.J., Neef, P., Caimi, M., Hoogland, T. and Claes, L.E., 'New in vivo measurements of pressures in the intervertebral disc in daily life', *Spine*, 15 April 1999; 24(8):753–62

Wognum, S., Huyghe, J.M. and Baaijens, F.P., 'Influence of osmotic pressure changes on the opening of existing cracks in 2 intervertebral disc models', *Spine*, 15 July 2006; 31(16):1783–8

Womersley, L. and May, S., 'Sitting posture of subjects with postural backache', *J Manipulative Physiol Ther*, March–April 2006; 29(3):213–18

Wood, K.B., Popp, C.A., Transfeldt, E.E. and Geissele, A.E., 'Radiographic evaluation of instability in spondylolisthesis', *Spine*, 1 August 1994; 19(15): 1697–703

Yang, K.H. and King, A.I., 'The mechanism of facet load transmission as a hypothesis for low-back pain', *Spine*, September 1984; 9(6):557–65

Yip, Y.B. and Tse, S.H., 'the effectiveness of relaxation acupoint stimulation and acupressure with aromatic lavender essential oil for non-specific low back pain in Hong Kong: a randomised controlled trial', *Complement Ther Med*, March 2004; 12(1):28–37

Zhao, F., Pollintine, P., Hole, B.D., Dolan, P. and Adams, M.A., 'Discogenic origins of spinal instability', *Spine*, 1 December 2005; 30(23):2621–30

Orders

Both the BackBlock (£28 plus £10 postage and packaging) and the MaRoller (£20 plus £5 postage and packaging) may be purchased from: Blenheim Estate Office, Blenheim Palace, WOODSTOCK, Oxfordshire, OX20 IPS, UK.

Please send a cheque for the items you wish to buy, along with your request. Please make cheques payable to Sunsar Blocks Limited.

Both come with instructions which must be strictly followed.